SpringerBriefs in Speech Technology

Studies in Speech Signal Processing, Natural Language
Understanding, and Machine Learning

Series Editor:
Amy Neustein

SpringerBriefs present concise summaries of cutting-edge research and practical applications across a wide spectrum of fields. Featuring compact volumes of 50 to 125 pages, the series covers a range of content from professional to academic. Typical topics might include:

- A timely report of state-of-the-art analytical techniques
- A bridge between new research results, as published in journal articles, and a contextual literature review
- A snapshot of a hot or emerging topic
- An in-depth case study or clinical example
- A presentation of core concepts that students must understand in order to make independent contributions

Briefs are characterized by fast, global electronic dissemination, standard publishing contracts, standardized manuscript preparation and formatting guidelines, and expedited production schedules.

The goal of the **SpringerBriefs in Speech Technology** series is to serve as an important reference guide for speech developers, system designers, speech engineers and other professionals in academia, government and the private sector. To accomplish this task, the series will showcase the latest findings in speech technology, ranging from a comparative analysis of contemporary methods of speech parameterization to recent advances in commercial deployment of spoken dialog systems.

More information about this series at http://www.springer.com/series/10043

Leena Mary • Deekshitha G

Searching Speech Databases

Features, Techniques and Evaluation Measures

Leena Mary
Department of Electronics
& Communication Engineering
Government Engineering College
Idukki, Kerala, India

Deekshitha G
Department of Electronics
and Communication Engineering
Rajiv Gandhi Institute of Technology
Kottayam, Kerala, India

ISSN 2191-737X   ISSN 2191-7388   (electronic)
SpringerBriefs in Speech Technology
ISBN 978-3-319-97760-7   ISBN 978-3-319-97761-4   (eBook)
https://doi.org/10.1007/978-3-319-97761-4

Library of Congress Control Number: 2018951796

Printed on acid-free paper

This Springer imprint is published by the registered company Springer Nature Switzerland AG
The registered company address is: Gewerbestrasse 11, 6330 Cham, Switzerland

Preface

In this new era, we deal with a large amount of multimedia data in our daily life. These multimedia files (many of which are audio and speech based) are archived for various applications. Such resources will be useful only if it is possible to retrieve the data efficiently whenever required. Hence, in relation to audio and speech, it is significant to have query-based audio search techniques to retrieve the needed speech data from a huge database.

Many algorithms and approaches are introduced for audio searching. Initially, the search was limited to a set of predefined keywords. But now the technologies have been developed to search for any spoken keyword. Different approaches have been introduced in the literature using different methodologies, datasets, features, and evaluation measures. This book *Searching Speech Databases: Features, Techniques, and Evaluation Measures* gives an overall idea about different audio search techniques attempted by researchers around the world.

The book describes the features and techniques related to audio search on speech databases. Chapter 1 gives a detailed introduction to audio search, its types, applications, evolution of techniques, databases, and platforms for benchmarking. Different evaluation measures are discussed in Chap. 2. Chapter 3 explains various features, their representations and matching techniques for searching speech databases. Chapters 4 and 5 describe different techniques for keyword spotting and spoken term detection, respectively.

The book is intended for system developers as well as the academic and research community.

Idukki, India — Leena Mary
Kottayam, India — Deekshitha G
June, 2018

Contents

Acronyms

ANN	Artificial Neural Network
ASM	Acoustic Segment Model
ASR	Automatic Speech Recognition
ATWV	Actual Term Weighted Value
BoAW	Bag of Acoustic Words
BUT	Brno University of Technology
CE-DTW	Constrained Endpoint Dynamic Time Warping
CNN	Convolutional Neural Network
DET	Detection Error Trade-off
DMLS	Dynamic Match Lattice Spotting
DNN	Deep Neural Network
DTW	Dynamic Time Warping
FA	False Alarm
FBCC	Fourier Bessel Cepstral Coefficient
FDLP	Frequency Domain Linear Prediction
GDS	Group Delay based Segmentation
GMM	Gaussian Mixture Model
GP	Gaussian Posteriorgram
HMM	Hidden Markov Model
HTK	HMM Toolkit
KWS	Keyword Spotting
LLR	Log Likelihood Ratio
LPCC	Linear Prediction Cepstral Coefficients
LVCSR	Large Vocabulary Continuous Speech Recognition
MED	Minimum Edit Distance
MFCC	Mel Frequency Cepstral Coefficients
MLP	Multilayer Perceptron
MTWV	Maximum Term Weighted Value
NIST	National Institute of Standards and Technology
NN	Neural Network

OOV	Out Of Vocabulary
OpenKWS	Open Keyword Search
PLP	Perceptual Linear Prediction
QbE	Query by Example
QUESST	Query by Example Search on Speech Task
ROC	Receiver Operating Characteristics
S-DTW	Segmental DTW
STD	Spoken Term Detection
SVM	Support Vector Machine
SWS	Spoken Web Search
TWV	Term Weighted Value
UE-DTW	Unconstrained Endpoints DTW
WER	Word Error Rate
ZCR	Zero Crossing Rate

Chapter 1
Audio Search Techniques

1.1 Introduction

With the advancement of communication technologies and Internet, the use of multi-media has increased exponentially. Most of the population switched to smart phones and are always making use of multimedia data. A large amount of multimedia data are freely available nowadays. But, any resource will become useful, only if we can handle them efficiently. In many gadgets, audio is considered as a data type. A huge amount of audio archives are available in different websites. Different audio archives include music databases, news bulletins, story databases, extempores, audio lectures, and audio interviews. Such audio resources can be efficiently used only if it is possible to retrieve the exact file from huge archives within a short span of time. Here lies the significance of audio search techniques. Audio search refers to search and retrieval of a particular audio file from an audio database.

With high speed–low cost internet connectivity, multimedia data has become a part of our daily life. Instead of typing a message, we prefer to send a voice message on handheld devices. Voice-guided applications are very popular nowadays. Hence, research on audio/speech is necessary to further improve the technologies.

Speech processing deals with speech signals and its processing. The main applications of speech processing include speech recognition, speaker recognition/verification, language recognition, audio search, speech enhancement, emotion recognition, speech coding, and speech synthesis. Audio search refers to the retrieval of desired audio data from audio archives. Speech data is a subclass of the audio data which includes the human speech data with a limited frequency range compared to music.

In this chapter, the task of audio search will be introduced, followed by the details of its classifications, applications, evolution, major milestones, different databases, and different benchmarking platforms.

© The Author(s), under exclusive licence to Springer Nature Switzerland AG 2019
Leena Mary, Deekshitha G., *Searching Speech Databases*, SpringerBriefs
in Speech Technology, https://doi.org/10.1007/978-3-319-97761-4_1

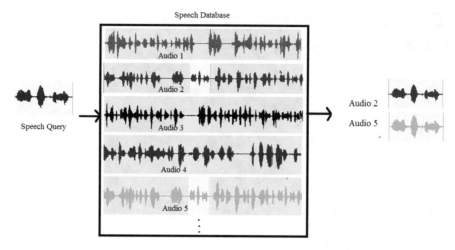

Fig. 1.1 Illustration of audio searching using speech query/keyword

1.2 Audio Search

Audio data implies all audible data in the frequency range of 20–20,000 Hz. It includes all audible sounds like speech data, music, animal sounds, bell sounds, laughter, bird chirps, news footage archive, and audio lectures. Such resources will be useful if and only if it is possible to find a particular file from database when needed, without much lapse of time. Since most of the audio archives are not well indexed or labeled, this is a challenging task. So researchers are working towards the development of better audio search algorithms.

With audio searching techniques, it is possible to retrieve a desired audio file by giving a keyword/query to the system as shown in Fig. 1.1. A keyword must be strong enough to uniquely describe the desired file from the database. Initial searching systems could work only with keyword in text form. But, recent systems have the ability to search with an audio query/keyword.

1.3 Classification of Audio Search Techniques

Audio searching techniques can be categorized according to different criteria such as type of audio, learning methods, and representation of query words. Audio search techniques are classified as shown in Fig. 1.2.

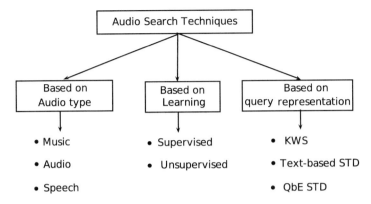

Fig. 1.2 Classification of audio search techniques

1.3.1   Classification Based on Audio Type

According to the type of audio data, the audio searching can be broadly classified into three categories namely: music, audio, and speech. For each audio type, the back-end approaches are different.

1.3.1.1   Music

Music is considered as a different category of audio, as music libraries differ very much compared to other audio types. For searching music files, many systems have been developed. It is the most popular and demanding task in audio search [1]. Several applications have been developed for searching music files just by singing/humming a part of it. Music retrieval becomes much more challenging due to the presence of background score.

1.3.1.2   Audio

Different types of audios (bell ringing, laughter, thunder, animal sounds, and water flow) are freely available in many online websites. People like sound engineers and composers are in need of such sound for their work. So, content-based classification of such sound categories as well as their retrieval is relevant. Many research works have been published in the area of content-based classification and retrieval [2, 4, 5].

Content-based audio classification refers to classification of audio recordings into categories like speech, music, environmental sounds, and silence. Content-based classification and retrieval systems use different techniques like fuzzy logic or features like temporal curves of energy function, average zero crossing rate, and fundamental frequency of audio signals. Classification and retrieval can be done using SVM classifier followed by a new metric called Distance from Boundaries (DFB) [4]. Audio retrieval is also done using semantic similarity [5]. Query-by Example (QbE) content-based audio data is retrieved by checking the semantic similarity rather than the acoustic similarity and this idea is illustrated using the *BBC Sound Effects library* [5].

1.3.1.3   Speech

Searching speech databases is a challenging task [51–72, 75–79, 82, 83]. Speech data can be of any language or mode (read/lecture/conversational). It is a challenging area, due to the variabilities in gender, age, dialect, and accent. Remaining part of this book describes various techniques developed for searching speech databases.

1.3.2   Classification Based on Type of Learning

Based on the learning techniques used, the search systems can be classified into supervised and unsupervised systems. Supervised systems perform well for the application or language for which it is trained. For under-resourced languages, supervised learning techniques are not feasible. Unsupervised learning techniques are mostly used in such cases.

1.3.2.1   Supervised

Systems that make use of transcribed audio files for learning can be categorized as supervised systems. Most of the searching systems that make use of Large Vocabulary Continuous Speech Recognition (LVCSR) or any speech to text conversion module can be classified as supervised systems. For building a supervised system, the effort needed is more, in terms of manually transcribing speech data. Manual transcription of speech files is a tedious job. Manually transcribed labels are used for training the models of supervised audio search systems.

1.3.2.2   Unsupervised

In the world, there are nearly about 7000 human spoken languages. Unfortunately, linguistic resources are not available for most of these languages. Commercially available ASR engines can support around 50–100 languages. For developing a search system for such under-resourced languages, large amount of transcribed data is required. Annotation is a time-consuming process which demands assistance of linguistic experts. Various unsupervised techniques are proposed by researchers for audio search [79, 82, 83].

1.3.3 Classification Based on Query Representation

Query is usually a keyword or a short phrase that is given to the system for retrieving an audio file containing that query. Based on the query type, the searching techniques can be broadly classified into two: keyword spotting and spoken term detection. Table 1.1 shows the classification of audio search techniques based on query representation.

1.3.3.1 Keyword Spotting

Keyword Spotting (KWS) is the foremost method in audio searching, where the system is designed for a set of predefined keywords [51–60]. Keywords are selected by considering the frequency of their occurrences. The words/phrases that are most frequently occurring are shortlisted as keywords. In KWS technique, speech to text conversion is done for both query and database. Text-level searching is done later. An LVCSR or its variants are usually employed for speech to text conversion. So, this method is useful only for well-resourced languages. Chapter 4 discusses more details about keyword spotting.

1.3.3.2 Spoken Term Detection

As the technologies evolved, it has become possible to search for any keywords from the audio archives. The query is entered as a spoken term or as a clipped audio file. The Spoken Term Detection (STD) systems can be classified into two: text-based STD and Query-by Example Spoken Term Detection (QbE-STD). Chapter 5 discusses about different STD approaches.

Text-Based STD

In text-based STD, the spoken query is first converted to corresponding text/symbols. Then, text-based search is done to find the occurrence of the keyword in the database [66–71]. ASR systems are required here for conversion of speech to text. As ASR systems need large amount of annotated speech data for their

Table 1.1 Classification of audio search techniques based on query representation

Method	Query type	Need for speech to text conversion
KWS	Predefined word/phrase	Yes
Text-based STD	Unspecific word/phrase	Yes
QbE-STD	Unspecific word/phrase	No

development, it cannot be used for under-resourced languages. For overcoming this limitation, QbE-STD was introduced.

Query-By-Example STD

In Query-By-Example (QbE) STD, unlike traditional STD the query will be processed directly [75–79, 82, 83]. There is no speech to text conversion in QbE-STD. User presents the system with desired audio snippets containing queries. The system then searches the database for segments that closely resemble these queries. Most of the QbE systems make use of template matching methods like DTW or its variants instead of ASR systems. Representation in terms of posteriorgrams is popular in QbE-STD.

1.4   Applications

Applications of audio search techniques include:

- Information retrieval
- Indexing and classifying of audio data
- Keyword spotting, monitoring, and retrieval
- Surveillance
- Music retrieval
- Monitoring television audio data for occurrence of a commercial
- Entertainment industry to play on demand
- Activation of some gadgets based on an audio input/instruction
- Consumer search

Audio archives like audio books, and classroom lectures can be indexed with the help of audio search algorithms. Keywords in the audio databases can be spotted out and monitored for many purposes. Similarly, searching techniques can be used for surveillance purposes like call center data monitoring. In the entertainment industry, many tools have been developed that make use of audio searching methods. For example, when a tune is hummed, the system will retrieve back a song similar to the input humming. Likewise in the near future, most of the devices will be designed to respond to oral instructions or audio tunes.

1.5   Audio Search: Software Packages and Patents

Most of the research in audio search are done on music database as it is the widely searched audio data type. Many software have been developed for searching music database and for content-based audio classification/retrieval. Some of the companies like "Shazam" and "Muscle Fish" work in this area.

Table 1.2 Some of the patents in the area of audio search

Publication No.	Title	Inventors	Year
EP 0 177 854 A1	Keyword recognition system using template-concatenation model	Alan Lawrence Higgins, Robert E. Wohlford, and Lawrence George Bahler	1985
US 7,058,575 B2	Integrating keyword spotting with graph decoder to improve the robustness of speech recognition	Guojun Zhou	2001
US 0,279,358 A1	Method and system for efficient spoken term detection using confusion networks	Brian E.D. Kingsbury, Hong-Kwang Kuo, Lidia Mangu and Hagen Soltau	2015
US 9,398,367,B1	Suspending noise cancellation using keyword spotting	Benjamin David Scott and Mark Edward Rafn	2016

Shazam [1] developed a commercial music search engine in which the algorithm used is highly noise/distortion resistant, computationally efficient, and fast. In Shazam mobile app, the system will quickly identify a short audio sample of music entered through the mobile's microphone, from a database containing millions of audio tracks. In Shazam browser, text-level search can also be done, so that the related audio files can be retrieved back.

Muscle Fish company [2] developed an application named "Sound Fisher" for content-based audio classification and retrieval. It is a sound-based database management system with content-based recognition and retrieval features. Various perceptual features such as loudness, brightness, pitch, and timbre are used to represent a sound. Neural Network (NN) is used for classification and retrieval. Database consists of different classes of sounds like laughter, animals, bells, etc. In the GUI, a query is formed using a combination of constraints like sound type, sampling rate, and date of release. By clicking the search button in the GUI, the search will be initiated and the results will be displayed.

Table 1.2 enlists some of the patents in the area of audio search. Convolutional Neural Network (CNN) based indoor navigation application named Halo has been proposed by [3], to accurately detect the orgin and destination locations from a spoken query utterance.

1.6 Evolution of Audio Search Techniques

Many audio search techniques have been introduced by researchers depending on the domain of application and performance requirement. An audio search system must ideally exhibit the following characteristics:

- Computational advantage
- Language independency

Table 1.3 Evolution of audio search techniques

Year	Technique
1973	Elastic template method
1978	Dynamic Time Warping (DTW)
1985	Continuous Speech Recognition + Template
1990	Hidden Markov Model (HMM) based
1991	Neural Network topologies
1993	Large Vocabulary Continuous Speech Recognition (LVCSR)
1994	Lattice-based approach
1995	Log-likelihood ratio scoring
1996	Indexing
1997	Predictive neural model
1998	Using active search algorithm
2005	Pattern discovery, Dynamic match phone lattice keyword spotting
2006	Posteriorgrams, Spoken Term Detection (STD)
2007	Minimum Edit Distance (MED), Confusion Networks
2008	Spectro-temporal patch features
2009	Segmental-DTW on Gaussian posteriorgrams, Query by Example (QbE), Fast phonetic decoding, phonetic posteriorgrams
2011	Segment-based DTW, Multilayer Perceptron (MLP), Acoustic Segment Model (ASM)
2012	Unconstrained Endpoints DTW, Group delay function, Spectrographic seam patterns
2014	Deep Neural Network (DNN), QbE using Frequency Domain Linear Prediction (FDLP) and Non Segmental DTW, Bag of Acoustic Words (BoAW)
2015	Fourier Bessel Cepstral Coefficients (FBCC), Convolutional Neural Network (CNN)
2016	Image processing techniques for STD
2017	Partial matching

- Fast response
- Less memory footage
- Higher accuracy
- Ability to handle text and audio queries

Table 1.3 shows the year-wise evolution or advancement in this area. Initially, the search was limited to a set of predefined keywords. But now, the technologies have been developed to search for any spoken keyword which is referred to as Query-by-Example Spoken Term Detection (QbE-STD).

1.7   Databases

Popular speech databases used by researchers for audio search are listed in Table 1.4 along with details such as language, type of data, and authorized link to the database. Among them NIST and MediaEval are benchmarking platforms, for which new languages are introduced every year.

Table 1.4 List databases for audio search

Sl. No.	Name	Language	Type	Link
1	TIMIT	American English	Microphone speech	[6]
2	2006 NIST	English, Mandarin Chinese, Standard Arabic, Arabic	Telephone conversations meeting speech, broadcast news	[7]
3	Indian languages	17 Indian languages	Read speech	[8]
4	International Corpus of English	English	Read speech	[9]
5	LWAZI corpus	African languages	Telephone channel	[10]
6	ELRA	European languages	Telephone, desktop and microphone Radio, Television, Internet Aurora	[11]
7	CALLHOME	Spanish	Telephone conversation	[12]
8	CSR-I (WSJ0) complete	English	Read speech	[13]
9	CSR-II (WSJ1) complete	English	Read speech	[14]
10	Fisher	English	Telephone conversations	[15]
11	MIT lecture corpus	English	Lecture	[16]
12	MediaEval	Indian, African languages	Read, telephone, elicited speech	[18]

1.8   Platforms for Benchmarking

Different approaches developed by researchers for the task of audio searching make use of different language databases and evaluation metrics. It is difficult to compare the performance of different systems. Thus, different platforms have emerged with an aim to benchmark the system performances. Most of them conduct yearly competitions to find out the best performing systems. They provide development and test database along with a standard metric to evaluate the systems in a uniform manner. Such platforms motivate the researchers working in this area to come up with innovative ideas. Some of the platforms are listed in Table 1.5.

Table 1.5 List of some benchmarking platforms for audio search systems

Evaluation series	Years of evaluation
NIST STD	2006
NIST (OpenKWS)	2013, 2014, 2015, 2016
Albayzin evaluation	2006, 2008, 2010, 2012, 2014, 2016
MediaEval (SWS)	2010, 2011, 2012, 2013
MediaEval (QUESST)	2014, 2015, 2016, 2017

1.8.1 NIST: Open Keyword Search Evaluation

The National Institute of Standards and Technology (NIST) started the Open Keyword Search (OpenKWS) [17] evaluation series in order to support research in this area. The goal of the program is to reduce the difficulty of building high-performing KWS systems on a new language quickly with limited data resources.

The OpenKWS project is an extension of the 2006 Spoken Term Detection evaluation, which tested KWS systems on English, Mandarin, and Arabic recordings from Broadcast News, Conversational Telephone Speech (CTS), and conference meeting data. OpenKWS is a multiyear effort to test KWS systems on a new "surprise language" each year. Participants will be supplied with conversational telephone speech for training, and testing the system, but had a limited time to build the systems. Results were then discussed in an evaluation workshop. Surprise languages were Vietnamese, Tamil, Swahili, and Georgian for 2013, 2014, 2015, and 2016, respectively. Metrics introduced by NIST for STD evaluation are Term Weighted Value (TWV), Actual Term Weighted Value (ATWV), and Maximum Term Weighted Value (MTWV) which will be discussed in the next chapter.

1.8.2 MediaEval

MediaEval is a benchmarking initiative dedicated to evaluate new algorithms for multimedia access and retrieval [18]. MediaEval attracts participants who are interested in multimodal approaches to multimedia involving, e.g., speech recognition, multimedia content analysis, and music and audio analysis.

1.8.2.1 Spoken Web Search Task

In 2011, 2012, and 2013, MediaEval organized Spoken Web Search (SWS) tasks [19]. The task involves searching for audio content within audio database using an audio query. This task is particularly interesting for speech researchers in the area of spoken term detection. A set of un-transcribed audio files from multiple languages and a set of queries will be provided in this task. The task addresses the challenge of multiple, resource-limited languages. It requires a language-independent audio search system, so that given a query, it should find the appropriate audio file and the location of query term.

For evaluation, the organizers followed the principles of NIST's Spoken Term Detection (STD) evaluations. In 2013, the participants were directed to report the running time of the system measured as the average real-time speed up obtained by the system to automatically search for a 1-s query term, in comparison to performing the search manually by listening to the whole reference data. Performance analysis of Spoken Web Search task of MediaEval 2011 and 2012 are available in the literature [20, 21].

1.8.2.2   QUESST

In 2014, SWS task got renamed to "Query by Example Search on Speech Task" (QUESST) [22]. The novelty of QUESST is in the nature of the queries being proposed as in Table 1.6. In addition to the single/multiword queries, complex single/multiword queries are also included in the task. Complexity in the queries is classified into three types: T1, T2, and T3 as shown in Table 1.6. The first type (T1) includes queries that might differ slightly (either at the beginning or at the end of the query) with exact match in the test utterance. The second type (T2) of complex queries corresponds to cases where two or more words in the query appear in different order in the search utterance. The third kind of complex queries (T3) is similar to the second one, but allowing the reference to contain some amount of "filler" content between the different matching words. For example, the query "white horse" should match with the utterances "My horse is white" as well as "I have a white and beautiful horse."

1.8.3   Albayzin Evaluation

This campaign is an internationally open set of evaluations supported by the Spanish Network of Speech Technologies and the International Speech Communication Association (ISCA) Special Interest Group on Iberian Languages, which have been

Table 1.6 Query types in QUESST

Query type	Example
T1	Research, Researcher
T2	White snow, Snow white
T3	White horse, My horse is white, I have a white and beautiful horse

held every 2 years since 2006. The evaluation campaigns provide an objective mechanism to compare different systems and are a powerful way to promote research on different speech technologies like query-by-example spoken term detection.

The Albayzin [23] search on speech evaluation involves searching in audio content for a list of terms/queries. This evaluation focuses on retrieving the appropriate audio files that contain any of those terms/queries. Two different tasks are defined: Spoken Term Detection (STD), where the input to the system is a list of terms, which is unknown while processing the audio. This is the same task as in NIST STD 2006 evaluation and Open Keyword Search in 2013, 2014, 2015, and 2016. The second task is Query-by-Example Spoken Term Detection (QbE STD), where the input to the system is an acoustic example per query and hence a prior knowledge of the correct word/phone transcription corresponding to each query cannot be made. This task must retrieve a set of occurrences for each query detected in the audio files, along with their time stamps as output, as in the STD task. QbE-STD is the same task as those proposed in MediaEval 2011, 2012, and 2013. Similarly, Intelligence Advanced Research Projects Activity (IARPA) recently completed the Babel program [24] intended on robust speech recognition and to support keyword search for any new language within a limited system build time of one week [25].

1.9 Summary

This chapter has discussed the basics, different classifications, and applications of audio search. The evolution of techniques was also discussed along with some leading databases and benchmarking platforms.

In the next chapter, details of different measures for evaluating the audio search systems will be discussed.

Chapter 2
Evaluation Measures for Audio Search

2.1 Introduction

This chapter gives the details of various metrics widely used for evaluating the performance of audio search techniques. Different audio search techniques can be compared using these measures. Earlier, the systems were evaluated using measures like accuracy, precision, recall, and likelihood scores. With the advent of NIST STD, metrics like Term Weighted Value (TWV), Actual Term Weighted Value (ATWV), and Maximum Term Weighted Value (MTWV) were introduced. These metrics got wide acceptance by the audio search community. Similarly, MediaEval platform introduced another metric named normalized cross entropy. These measures are explained in this chapter.

2.2 Evaluation Measures

2.2.1 Accuracy and Error Rate

Accuracy is a simple and mostly used measure to evaluate a system. It is the ratio between the number of items correctly classified to the total number of items. Table 2.1 indicates four types of classifications.

"True Positive (TP)" refers to the instances that are correctly classified as the class of interest, "True Negative (TN)" indicates the instances that are correctly classified as uninterested class. Similarly, "False Positive (FP)" refers to the instances that are incorrectly classified as the class of interest, and "False Negative (FN)" indicates the instances that are wrongly classified as uninterested class.

$$Accuracy = \frac{TP + TN}{TP + TN + FP + FN} \tag{2.1}$$

© The Author(s), under exclusive licence to Springer Nature Switzerland AG 2019
Leena Mary, Deekshitha G., *Searching Speech Databases*, SpringerBriefs
in Speech Technology, https://doi.org/10.1007/978-3-319-97761-4_2

Table 2.1 Accuracy and error rate

		Actual class	
		Positive	Negative
Predicted	Positive	True Positive (TP)	False Positive (FP)
class	Negative	False Negative (FN)	True Negative (TN)

The error rate of the system is calculated as:

$$Error\ rate = 1 - Accuracy = \frac{FP + FN}{TP + TN + FP + FN} \qquad (2.2)$$

The desired value of accuracy and error rate are 1 and 0, respectively.

2.2.2 Sensitivity and Specificity

Sensitivity gives a measure of actual-true data. Sensitivity measures the proportion of positive examples that are correctly classified. This is also known as "True-positive value."

$$Sensitivity = \frac{TP}{TP + FN} \qquad (2.3)$$

Specificity gives a measure of actual-false data. Specificity measures the proportion of negative examples that are correctly classified. This is also known as "True-negative value."

$$Specificity = \frac{TN}{TN + FP} \qquad (2.4)$$

Sensitivity and specificity range from 0 to 1, with values close to 1 being more desirable.

2.2.3 Precision and Recall

Precision gives the confidence measure of a system. It gives the amount of positive predictions. It is the proportion of positive examples that are truly positive. It shows the fraction of retrieved instances that are relevant to the query.

$$Precision = \frac{TP}{TP + FP} \qquad (2.5)$$

Precision takes all retrieved instances into account, but it can also be evaluated at a given cut-off rank, considering only the N top most results returned by the system. This measure is called P at N ($P@N$) [26, 82].

Recall is a measure similar to sensitivity. Recall in information retrieval is the fraction of relevant instances to the query that are successfully retrieved [82].

$$Recall = Sensitivity = \frac{TP}{TP + FN} \tag{2.6}$$

2.2.3.1 Mean Average Precision

Mean Average Precision (MAP) which is the mean of the precision scores after each query hit is retrieved. It provides a single-figure measure of quality across the recall levels. Among evaluation measures, MAP has shown good discrimination and stability. Average precision is the average of the precision value obtained for the set of top k documents existing after each relevant document retrieval, and this value is then averaged over information needs. That is, if the set of relevant documents for a query $q_j \in Q$ is $d_1, \cdots, d_{m_j}$ and R_{jk} is the set of ranked retrieval results from the top result until the document d_k is obtained, then

$$MAP(Q) = \frac{1}{|Q|} \sum_{j=1}^{|Q|} \frac{1}{m_j} \sum_{k=1}^{m_j} Precision(R_{jk}) \tag{2.7}$$

When a relevant document is not retrieved at all, the precision value in the Eq. (2.7) is taken to be 0 [27]. Or, the average precision is approximated as the area under the uninterpolated precision–recall curve, and hence the MAP is roughly the average area under the precision–recall curve for a set of queries [82].

2.2.4 Miss Rate and False Alarm Rate

Miss rate or false-negative rate gives the measure of actual instances that are missed while predicting.

$$Miss\ Rate = \frac{FN}{FN + TP} = (1 - True\ positive\ Rate) \tag{2.8}$$

False alarm rate or fall-out gives the measure of incorrectness in the prediction of a system. It is also known as "False-positive rate."

$$False\ alarm\ rate = 1 - Specificity = \frac{FP}{FP + TN} \tag{2.9}$$

2.2.5 F-Measure

F-Measure combines the value of precision and recall to a single number using harmonic mean. Harmonic mean is used rather than the more common arithmetic mean, since both precision and recall are expressed as proportions between zero and one.

$$F - Measure = \frac{2 \times Precision \times Recall}{Recall + Precision} = \frac{2 \times TP}{2TP + FP + FN} \qquad (2.10)$$

2.2.6 Receiver Operating Characteristics

The Receiver Operating Characteristics (ROC) curve is commonly used to examine the trade-off between the detection of true positives, while avoiding the false positives. It is a visual measuring unit so that the system's performance for a wide range of values can be observed from a single curve. It is a plot with False-Positive Rate (FPR) on X-axis and True-Positive Rate (TPR) on Y-axis. FPR is equivalent to *(1-specificity)* and TPR is equivalent to *sensitivity*. A sample ROC characteristic is shown below in Fig. 2.1.

In ROC plot, the closer the curve to the perfect classifier, the better is the system in identifying true-positive values. This can be measured using another metric named "*Area under ROC.*" For this, ROC curve is considered as a two-dimensional

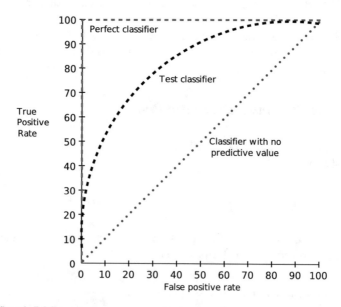

Fig. 2.1 Sample ROC curve

Table 2.2 Values of area under ROC and its interpretation

Area under ROC	Meaning
1	Perfect prediction
0.9	Excellent prediction
0.8	Good prediction
0.7	Mediocre prediction
0.6	Poor prediction
0.5	Random prediction
<0.5	Something wrong

plot for calculating area under the ROC curve. If the area is closer to 1, it shows perfect prediction. See Table 2.2 to understand the meaning of different values of *Area under ROC*.

2.2.7 Figure of Merit

The Figure of Merit (FOM) is a well-established evaluation metric used for the KWS task [55]. In order to summarize the ROC curve to a single value, the FOM is used. It is the detection rate for target patterns when the false alarm rate is over a specified range.

The FOM is defined as the detection rate averaged over the range of 0–10 false alarms per hour and over the individual queries (fa/kw-h). Equivalently, it can be interpreted as the normalized area under the ROC curve in that false alarm range. The log-likelihood scores and word boundary information from the output of a speech recognizer can be used to compute the FOM. FOM is defined by NIST as an upper-bound estimate on word spotting accuracy averaged over 1–10 false alarms per hour. The FOM is calculated as follows: it is assumed that the total duration of the test speech is T hours. For each word, all of the spots are ranked in score order. The percentage of true hits p_i found before the ith false alarm is then calculated for $i = 1 \cdots , N + 1$ where N is the first integer $\geq 10T - 0.5$. Then, the figure of merit is defined as:

$$FOM = \frac{1}{10T}(p_1 + p_2 + \cdots + a.p_{N+1}) \qquad (2.11)$$

where $a = 10T - N$ is a factor that interpolates to 10 false alarms per hour [28].

2.2.8 Detection Error Trade-Off

Like ROC, Detection Error Trade-off (DET) is another visual measure for evaluating the performance of a system. DET plot has False-Positive Rate (FPR) on X-axis and

Fig. 2.2 Sample DET curve

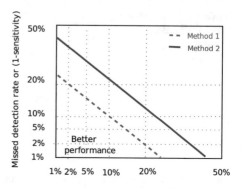

Miss Rate on the Y-axis, where miss rate is equivalent to (1-sensitivity). X and Y axes are scaled nonlinearly by their standard normal deviates (or just by logarithmic transformation) yielding trade-off curves that are more linear than ROC curves. A sample DET curve is shown below in Fig. 2.2.

2.2.9 Term Weighted Value

A system's "*value*" measures the usefulness of that system. A perfect system always responds correctly to a stimulus; however, an omitted response or a misleading response reduces the "value" of a system to a user. Thus, Term-Weighted Value (TWV) is one minus the average value lost by the system per term [29, 66]. The value lost by the system is a weighted linear combination of P_{Miss} and P_{FA}, where P_{Miss} and P_{FA} denote the miss probability and false alarm probability, respectively. The weight, β, takes into account both the prior probability of a term and the relative weights for each error type.

$$TWV(\theta) = 1 - average_{term}\{P_{Miss}(term, \theta) + \beta.P_{FA}(term, \theta)\} \qquad (2.12)$$

where

$$\beta = \frac{C}{V}(Pr_{term}^{-1} - 1)$$

θ is the detection threshold, $P_r(term)$ is the prior probability of a term, and C/V is the cost/value ratio, which is the ratio of value lost by a false alarm to the value lost by a miss. Miss and false alarm probabilities for a given query "term" are functions of the detection threshold θ:

$$P_{Miss}(term, \theta) = 1 - \frac{N_{correct}(term, \theta)}{N_{true}(term)}$$

$$P_{FA}(term, \theta) = 1 - \frac{N_{spurious}(term, \theta)}{N_{NT}(term)}$$

where:

- $N_{correct}(term, \theta)$ is the number of correct detections (retrieved by the system) of the query "$term$" with a score greater than or equal to θ.
- $N_{spurious}(term, \theta)$ is the number of spurious detections of the query "$term$" with a score greater than or equal to θ.
- $N_{true}(term)$ is the number of true occurrences of the query "$term$" in the corpus.
- $N_{NT}(term)$ is the number of opportunities for incorrect detection of the query "$term$" in the corpus; it is the "Non-Target" query trials. It has been defined by the following formula: $N_{NT}(term) = T_{speech} - N_{true}(term)$. T_{speech} is the total amount of speech in the collection (in seconds).

A detection threshold θ has been determined empirically per source type and the hard decision of the occurrences having a score less than θ is set to false; false occurrences returned by the system are not considered as retrieved and, therefore, are not used for computing.

The maximum possible TWV is 1.0, corresponding to "perfect" system output: no misses and no false alarms. The TWV of a system that outputs nothing is 0 and negative values of TWVs are also possible.

2.2.10   Actual vs. Maximum Term Weighted Value

While DET curves represent performance for all possible values of θ, two points on the DET curve are of interest because they determine if the system's actual decision threshold is optimal or not. The first is Actual Term-Weighted Value (ATWV), which is the TWV using the actual decisions [29]. ATWV represents the system's ability to predict the optimal operating point given the TWV scoring metric. It ranges from $-\infty$ to $+1$.

The second is Maximum Term-Weighted Value (MTWV). MTWV is the TWV at the point on the DET curve where a value of θ yields the maximum TWV. It ranges from 0 to $+1$. The difference between the values for ATWV and MTWV indicates the benefit of selecting a better actual decision threshold θ. The TWV, ATWV, and MTWV are the evaluation measures defined by NIST for the 2006 STD evaluation [29].

2.2.11 Word Error Rate

Word Error Rate (WER) is a common metric for analyzing the performance of a speech recognition system. Word Error Rate (WER) is used to characterize the accuracy of the transcripts. *WER* is defined as follows:

$$WER = \frac{S + D + I}{N} \times 100 \qquad (2.13)$$

where N is the total number of words in the corpus, and S, I, and D are the total number of substitution, insertion, and deletion errors, respectively. The Substitution Error Rate (SUBR) is defined by:

$$SUBR = \frac{S}{S + D + I} \times 100 \qquad (2.14)$$

Deletion Error Rate (DELR) and Insertion Error Rate (INSR) are defined in a similar manner.

Word Accuracy (W_{Acc}) is defined as:

$$W_{Acc} = 1 - WER = \frac{N - S - D - I}{N} \qquad (2.15)$$

Word Correctness (W_{Corr}) is defined as:

$$W_{Corr} = \frac{N - S - D}{N} \qquad (2.16)$$

2.2.12 Normalized Cross Entropy Cost

The normalized cross entropy cost (C_{nxe}) [30] was introduced by QUESST 14 (see Sect. 1.8.2.2) as a primary metric for evaluation. The C_{nxe} was introduced as an attempt to evaluate whether such a metric can be used as a feasible alternative to the ATWV metric, which has received many criticisms over the years, due to the embedded working point decisions.

Moreover in TWV scoring, the cost of missing a hit depends on the number of true occurrences of the query in the data set. On the other hand, false alarm is equally expensive for less as well as widely occurring queries. The global TWV is averaged over each query's TWV, so that each query has equal weight. TWV "forces" to lower threshold for less occurring queries. It is better to "pay a bit" for several false alarms than to "pay a lot" for one miss (especially for less occurring queries). This leads to a dependency of the threshold on the number of the true query occurrences.

System performance can be evaluated in terms of C_{nxe}, which is only based on system scores, in contrast to TWV. C_{nxe} measures the fraction of information, with regard to the ground truth, that is not provided by the system scores, assuming that they can be interpreted as log-likelihood ratios (LLR). A perfect system would get $C_{nxe} = 0$ and a non-informative (but well-calibrated) system would get $C_{nxe} = 1$, whereas $C_{nxe} > 1$ would indicate severe mis-calibration of the log-likelihood ratio scores [31].

2.3 Summary

Different performance measures used for evaluating audio search techniques are discussed in this chapter. A brief idea on the value of those measures for a perfect system is also covered. Details of graphical performance evaluation metrics like DET and ROC are discussed.

In the next chapter, different feature sets and their representations which are shown to be effective for audio search will be explained along with some matching techniques.

Chapter 3
Features, Representations, and Matching Techniques for Audio Search

3.1 Introduction

Feature extraction is very important in pattern recognition tasks, as it reduces the dimensionality and hence the computation complexity. In this chapter, different primary features popular for audio search are discussed. Different representations used for audio searching such as posteriorgrams are also discussed. Such representations can be realized with the help of Neural Networks (NN) or Gaussian Mixture Models (GMMs). Template matching methods like Dynamic Time Warping (DTW) and its successors are also important in audio searching. A detailed discussion is presented on template matching, variants of DTW and Minimum Edit Distance (MED) measures.

3.2 Primary Features

Most important cues from the signal can be parametrized by feature extraction. Ideally, the features chosen must be robust against any variations. Features directly extracted from speech are referred to as primary features. Features like MFCC, LPCC, FBCC, etc., come under the category of primary features. For extracting such features, mostly the speech signal is segmented into short frames of 20–30 ms, with an overlap of 10 ms typically.

3.2.1 Mel-Frequency Cepstral Coefficients

Mel-Frequency Cepstral Coefficients (MFCC) is considered to be the best available approximation of human perceptual characteristics [32]. Mel-Frequency Cep-

Leena Mary, Deekshitha G., *Searching Speech Databases*, SpringerBriefs in Speech Technology, https://doi.org/10.1007/978-3-319-97761-4_3

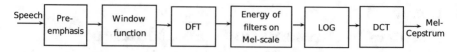

Fig. 3.1 Block diagram of MFCC extraction

strum (MFC) is a representation of the short-term power spectrum of speech, based on Discrete Cosine Transform (DCT) of log power spectrum on nonlinear Mel-scale of frequency. It is observed that human ears act as a filter and it concentrates on certain frequency components. These filters are nonuniformly spaced on the frequency axis, with more filters in the low frequency regions and less number of filters in the high frequency regions. This nonuniformly spaced filters are termed as Mel-filters. Cepstral coefficients obtained on Mel-spectrum are referred to as Mel-Frequency Cepstral Coefficients. The basic block diagram for MFCC extraction is shown in Fig. 3.1. For extracting MFCC features from speech signal, first the signal is analyzed over short analysis window. For each short window of speech, a spectrum is obtained using DFT. Then, this spectrum is passed through Mel-filters to obtain Mel-spectrum followed by a logarithmic transformation and DCT. Since we have performed DFT, DCT transforms the frequency domain into a time-like domain called *quefrency* domain. Cepstral analysis is performed on Mel-spectrum to obtain features that are similar to cepstrum, and hence termed as Mel-Frequency Cepstral Coefficients (MFCCs). Thus, speech is represented as a sequence of cepstral vectors. Cepstral analysis like liftering is done to separate the vocal tract and excitation characteristics in the quefrency domain.

Usually, each frame is converted into 12 MFCCs plus a normalized energy parameter. The initial coefficient C_0 represents the average energy in the speech frame and is often discarded for amplitude normalization. Estimation of first and second derivatives of MFCCs, along with energy, results in a 39-dimensional feature vector for each speech frame.

3.2.2   Fourier–Bessel Cepstral Coefficients

The Fourier series representation employs an orthogonal set of sinusoidal functions as a basis set, while the Fourier transform uses complex exponential functions. The sinusoidal functions are periodic and are ideal for representing general periodic functions. But, in case of speech signal, which is quasi-stationary, these are not efficient. Based on this, several aperiodic non-sinusoidal functions including exponentially modulated sinusoid and Bessel functions have been used for speech analysis with varying degree of success. In the Fourier–Bessel transformation, the basis function is a damped sinusoid which is more natural for the representation of voiced speech signal. Therefore, several experiments are being conducted using Fourier–Bessel Cepstral Coefficients (FBCC) in the areas of speech recognition, speaker identification, and QbE-STD.

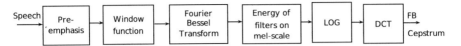

Fig. 3.2 Block diagram of FBCC extraction

In FBCC, instead of Fourier transform, Fourier–Bessel expansion is used for getting the frequency domain representation of the signal. Generally, vocal tract is modeled as a set of linear acoustic tubes, being cylindrical in shape. It can be efficiently modeled using FB expansion because Bessel functions are solutions to cylindrical wave equations. Representing speech signals using FB coefficients helps in identifying different components present in the signal [33]. Figure 3.2 shows the block diagram of FBCC extraction. The Fourier–Bessel transform is applied to the framed speech. Remaining operations are the same as that of MFCC extraction.

3.2.3 Linear Prediction Cepstral Coefficients

In Linear Prediction (LP) analysis, an all-pole model is used for the approximation of vocal tract spectral envelope.

A speech sample can be predicted by the linear weighted sum of past p samples, where p is the order of prediction [32–35].

$$\hat{s}(n) = -\sum_{k=1}^{p} a_k.s(n-k) \tag{3.1}$$

where $\hat{s}$ is the predicted speech sample, s is the windowed speech signal given by $s(n) = x(n).w(n)$, and a_k's are the LP coefficients. Then, the prediction error is calculated as:

$$e(n) = s(n) - \hat{s}(n) = s(n) + \sum_{k=1}^{p} a_k.s(n-k) \tag{3.2}$$

$$Total\ prediction\ error = E_n = \sum_{m} e_n^2(m) = \sum_{m} \left[S_n(m) - \sum_{k=1}^{p} a_k S_n(m-k) \right]^2 \tag{3.3}$$

where E_n is the error signal at time n, $S_n(m) = S(n+m)$ is the short-term speech segment, and $e_n(m) = e(n+m)$ is the short-term speech segment. To calculate the predictor coefficients, differentiate E_n with respect to each a_k:

$$\frac{\partial E_n}{\partial a_k} = 0,\ k = 1, 2, \cdots, p \tag{3.4}$$

to get

$$\sum_m S_n(m-i)S_n(m) = \sum_{k=1}^{p} a_k \sum_m S_n(m-i)S_n(m-k). \tag{3.5}$$

In least square terminology, the set of equations in Eq. (3.5) are termed as *normal equations*. From this set of p equations, p predictor coefficients can be solved out, which minimizes E_n. Depending on the range over which the error is minimized, there are two distinct methods for estimating these $a'_k s$ [34].

1. Autocorrelation method: Here, the error is minimized over the infinite duration $-\infty < m < \infty$.

$$\sum_{k=1}^{p} a_k R(i-k) = -R(i), \ 1 \leq i \leq p \tag{3.6}$$

where $R(i) = \sum_{m=-\infty}^{\infty} S_n(m)S_n(m+i)$ is the autocorrelation function of the signal.
2. Covariance method: In contrast to the autocorrelation method, here the error is minimized over a finite interval ($0 \leq m \leq N-1$).

A very useful parameter set that can be derived from the LP coefficients is the LP Cepstral Coefficients (LPCC) [35]. It seems to be more robust and reliable feature set for speech recognition than the LP coefficients. The recursion used for the computation of LPCC is:

$$c_0 = \ln \sigma^2 \tag{3.7}$$

$$c_m = a_m + \sum_{k=1}^{m-1} \left\{\frac{k}{m}\right\} c_k a_{m-k}, \ 1 \leq m \leq p \tag{3.8}$$

$$c_m = \sum_{k=1}^{m-1} \left\{\frac{k}{m}\right\} c_k a_{m-k}, \ m > p \tag{3.9}$$

where σ^2 is the gain term in the LP analysis.

3.2.4 Perceptual Linear Prediction Cepstral Coefficients

Compared to LP analysis, Perceptual Linear Prediction (PLP) analysis is more consistent with human hearing [36]. In PLP analysis, the power spectrum of the speech signal is modified before doing LP all-pole modeling. It makes use of psychophysics of hearing to derive an estimate of auditory spectrum. Mainly, three concepts included are: (1) the critical band spectral resolution, (2) the equal loudness

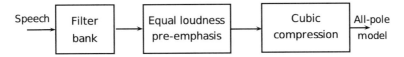

Fig. 3.3 Block diagram for modifying the power spectrum in PLP analysis

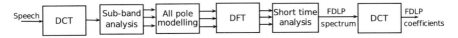

Fig. 3.4 Block diagram of FDLP extraction

curve, and (3) the intensity loudness power law. Figure 3.3 shows the block diagram for modifying the power spectrum in PLP analysis. Later sections remain the same as that of LP analysis.

3.2.5 Frequency Domain Linear Prediction Cepstral Coefficients

In noisy environments, Frequency Domain Linear Prediction (FDLP) features perform better than other short-time spectral features [37]. Figure 3.4 shows the block diagram for extracting FDLP coefficients.

3.3 Feature Representation for Audio Search

Instead of using the primary features directly, they are represented in different forms for the keyword spotting task. These representations are usually derived from the primary features. Such secondary features or representations popular in the area of audio search are discussed here.

3.3.1 Seam Patterns

The concept of seams has been adapted from the area of image processing. Seam carving is an image processing technique for resizing images without affecting the aspects of their content. With the help of a spectrogram, an audio can be conveniently represented as an image. A set of connected pixels in a spectrographic image is termed as a path or trajectory. A seam is defined as a trajectory that has the maximum line-integral value or has patterns of high-energy tracks [38]. Seam represents the path of maximum spectral "whiteness" across frequency in spectrogram. For similar type of sound, seams will be least variant/almost invariant. But for different sounds, seams will be different.

Seam is computed using a dynamic programming technique. Consider a position (i, j) in the 2-D matrix of the spectrogram, where i denotes the row index and j denotes the column index. For computing seam, three parameters are needed: (1) $E[i, j]$: energy element, (2) $C[i, j]$: cumulative energy leading up-to that element, and (3) $P[i, j]$: the path matrix which directs to previous element's index.

$$C[i, j] = E[i, j] + \max_{k \in [j-2, j+2]} C[i-1, k] \tag{3.10}$$

The maximum value of the last row of matrix C indicates the end of the maximum energy containing seam. From this maximum cumulative energy cell, back training is done to find the required seam as shown in Fig. 3.5. Since the spectrogram is much coarser in information, the seams obtained from them are usually jagged/rough. In order to smooth out, basic smoothing filter with a linear penalty is used for computing the seam pattern as:

$$C[i, j] = E[i, j] + \max_{k \in [j-2, j+2]} Pen[i, k].d.C[i-1, k] \tag{3.11}$$

where d is deviation distance and $Pen[i, k]$ is a penalty factor that depends on $(i - k)$. The penalty smooths out the seams and forces them not to change tracks unnecessarily. Seams are time invariant. To enforce uniformity to seams obtained from different windows, an origin is chosen. An approach similar to finding center of mass of a given body is used, to ensure that the number of elements of a seam is equally distributed on the two sides.

Individual seams can vary from instance to instance and are not individually useful. But, the ensemble that carries information are relevant for classification. Hough transform (a classic feature extraction technique) is used to capture the characteristics of this ensemble. Hough transform is computed from all the points on all detected seams and makes two adjustments. First, the origin of the transform is taken to lie at zero frequency. Then, since most regions in the transform are relatively low in intensity and contain little information, only the central region of the transform is retained. This retained central portion of the Hough transform is a matrix of numbers, which is resolved into a "Seam–Hough" feature vector.

Fig. 3.5 Seams for the word *kindergarten* (from sx101 of TIMIT database)

3.3.2   Patches

Most word-spotting algorithms depend on MFCC as features, where MFCC provides only the spectral information. Temporal sensitivity can be added to MFCC features through their deltas and delta-deltas. Moreover, MFCC is a *global* feature where the parameters span the entire frequency range. To overcome these constraints of MFCC, spectro-temporal patches are suggested [39].

It is being proved that keyword spotting task can be performed well using these spectro-temporal patch features, as it can provide spectral as well as temporal sensitivity. Patch features are extracted from the Mel-spectrogram of the speech signal [39]. In spectrograms, different characteristics such as formants, horizontal lines related to the pitch of the speaker, and vertical on–off edges in time can be observed. Unlike MFCC, the characteristics in spectrogram are said to be *local*, since the formant, harmonic, or noise pattern do not affect the entire spectrogram but seem to occur in local areas. Moreover, studies show that the patch-like features are more selective than frame-like features to the auditory cortical cells. Thus, 2D spectro-temporal patches are suggested as a feature set for keyword spotting. The patches are extracted from random locations of the keyword. The width and height of the patches are kept uniform for a spectral range of F_{range} and temporal range of T_{range}. Patch feature $P_k(f, t)$ is considered as a matched filter to find similar patterns. Since speech is a temporal pattern, it is necessary to keep the order of the patches occurring in the keyword. For that, patch location is recorded in terms of frequency f_k and relative time rt_k. Thus, a patch dictionary is created consisting of K patches $\{P\}_{k=1}^{K}$, their center location in frequency $\{f\}_{k=1}^{K}$, and relative time $\{rt\}_{k=1}^{K}$. This arranged dictionary is termed as *ordered-bag of spectro-temporal patches*. Figure 3.6 shows the Mel-spectrogram and sample patch extraction from the word *kindergarten* for sx101 of TIMIT database.

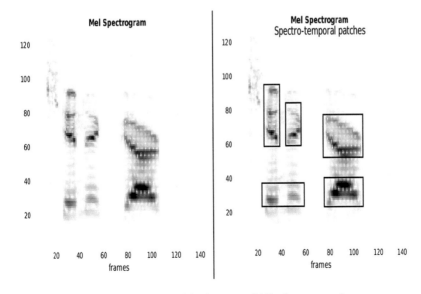

Fig. 3.6 (**a**) Mel-spectrogram of the word *kindergarten*. (**b**) Patches extracted

3.3.3 Posteriorgrams

Posterior features have been widely used in template-based speech recognition system. Gaussian posteriorgrams [79] and phonetic posteriorgrams [77] are the main posteriorgram representations. Primary features such as MFCC/FBCC are used for obtaining posteriorgrams.

Phonetic posteriorgram is a time vs class matrix, representing the posterior probability of each phone class for each time frame. Figure 3.7 shows a sample posteriorgram representation. Horizontal axis represents the time and vertical axis represents the phonetic classes. To generate a phonetic posteriorgram, a phonetic classifier is needed and hence it is a supervised technique. Each matrix entry represents the posterior probability, with black color showing lower probabilities and white color indicating highest values.

To deal with an under-resourced language, Gaussian posteriorgrams are used. Speech is directly modeled using a single GMM without any supervision. Generally, the number of Gaussian components must match with the number of broad phone classes in the language under consideration. If the number of components is small, the GMM training will result in under-fitting problem, where the detection rate reduces. If the number of components is large, the model will become very sensitive to variations in the training data, so that it causes generalization errors on the test data. The model is characterized by the value of their parameters. The parameters are mean, variance, and the weights. The initial value for the mean can be obtained by using K-means algorithm. Each Gaussian component approximates a phone-like class or a syllable class. By calculating the posterior probability for each frame on each Gaussian component, a posterior feature representation called a Gaussian posteriorgram (GP) is obtained. In GP representation, horizontal axis represents time whereas vertical axis shows the Gaussian cluster index. Consider a speech utterance with n frames as $S = (s_1, s_2, \cdots, s_i, \cdots, s_n)$. Then, the Gaussian posteriorgram of that speech utterance is given by $GP(S) = (q_1, q_2, \cdots, q_i, \cdots, q_n)$, where vector q_i is defined as:

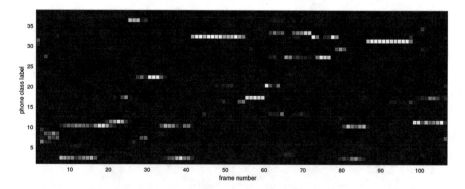

Fig. 3.7 An example of phonetic posteriorgram representation for the spoken phrase "*barometric pressure*"

$$q_i = (P(C_1/s_i), P(C_2/s_i), \cdots, P(C_m/s_i)) \tag{3.12}$$

where C_i represents the ith Gaussian component of GMM and m denotes the number of Gaussian components.

Another type of posteriorgrams used is Acoustic Segment Model (ASM) posteriorgrams. Similar to GMM posteriorgrams, ASM posteriorgrams can be trained in an unsupervised manner. In GMM training, each speech frame is considered to be independent. But in ASM training, similar neighboring frames are grouped together into smaller segments and then models are established for each of these segments. Since grouping is done, ASM makes use of temporal information of speech which is discarded in GMMs. Moreover, ASM-based speech recognizer is similar to conventional phoneme recognizer, where each sound unit is modeled by an HMM rather than a single Gaussian. The unsupervised iterative ASM training consists of mainly three steps: initial segmentation, segment labeling, and HMM training, which results in a set of phoneme-like ASMs. In [82], Euclidean distance is used as the distortion metric for segmentation and agglomerative clustering method is adopted. The ASM posteriorgrams reflect the posterior distribution over a set of ASM units.

3.3.4   Lattice of Segment Labels

Lattice is a loop-free directed graph, in which each node of the graph is associated with a time point of the utterance along with a label showing the phone hypothesis [40]. The likelihood score of the phone hypothesis is also represented at each edge of the lattice. Thus, a lattice can store the details of multiple phone sequence given by the continuous speech recognizer. Usually, a recognizer outputs the best phone hypothesis by using the Viterbi algorithm. But in the lattice, at every speech frame, the top N phone hypothesis ending at that node are stored. Here, N refers to the *degree* or *depth* of the lattice. A lattice structure with degree 1 gives the phone sequence with maximum likelihood probability. Since a lattice can handle N best phone sequence, out-of-vocabulary issue can be solved. A sample section of the lattice of degree 2 containing the word *bedroom* is shown in Fig. 3.8.

Similar to phone lattices, word/syllable/hybrid (with words and sub-word units) lattices can also be constructed. Word lattice can provide accurate term detection than phone lattices but the LVCSR decoding will be slow, resulting in slower

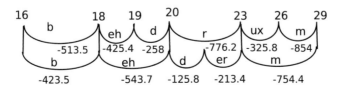

Fig. 3.8 Section of the lattice of degree 2 containing the word *bedroom*

indexing speed. Even though word lattice-based searching system requires large index space and slow indexing, the searching will be done instantaneously. In phone lattice-based systems, the index size remains more or less the same, and indexing is comparatively faster, but searching will be slower. The major drawbacks of the word lattice-based approach are: (1) its inability to detect OOV words and (2) recognition errors, as the ASR language model tries to give meaningful sentences as output [68], (3) slow decoding of LVCSR, (4) need of large amount of data to train the LVCSR, (5) increase in the computational cost, and (6) the system trained might be insufficient for data in another domain of interest [69]. On the other hand, the main advantages of using phone lattice is that: (1) it can avoid OOV problems, (2) no need for lexicon or pronunciation dictionary, (3) can be made language independent with a phone recognizer having rich phone set, and (4) can reduce the computational demands than using an ASR system [66, 68–70].

However, the use of sub-word lattices will increase the size of index table and is expensive in time to build the index table and need more space to store them [43]. Efforts to increase the speed of LVCSR decoding resulted in increased WER, and degraded STD performance [71]. Moreover, fusion of word and phonetic indices were also tried to improve the STD accuracy, so that word-level index can be used for detecting in-vocabulary terms and phonetic index for the OOV words [41–44].

3.3.4.1   Confusion Network

A confusion network is the most *compact representation* of multiple hypotheses of the original lattice structure. All the word/sub-word hypotheses are totally ordered in this confusion network and it helps in error minimization. This network is more useful to do keyword spotting of OOV words, as it has more paths than the original lattice [43]. Confusion network is an excellent and precise representation of lattice if one aims to narrow down the large search space associated with speech recognition to a small set of mutually exclusive, single word/sub-word entries.

The alignment shown in Fig. 3.9b is equivalent to the lattice and is referred to as a confusion network. The confusion network has one node for each equivalence class of original lattice nodes (plus one initial/final node), and adjacent nodes are linked by one edge per word hypothesis (including the null word) [45].

3.3.5   Bag of Acoustic Words and Inverted Indexing

The bag-of-words model is a simplified representation used in natural language processing and text retrieval systems. A text can be represented as a bag of its words, irrespective of grammar or even word order in the text, but keeping the multiplicity. Bag of Acoustic Words (BoAW) [78] got inspired from the concept of Bag of Words, and proposed to be useful for audio search applications. In BoAW, a speech segment

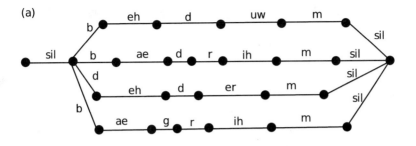

(a)

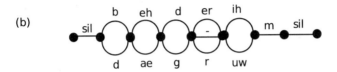

(b)

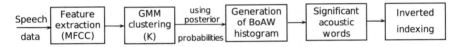

Fig. 3.9 Lattice vs. confusion network. (**a**) Lattice structure for the word *bedroom*. (**b**) corresponding confusion network where "-" indicates deletions

Speech data →	Feature extraction (MFCC)	→	GMM clustering (K)	using posterior probabilities	Generation of BoAW histogram	→	Significant acoustic words	→	Inverted indexing

Fig. 3.10 Block diagram of BoAW representation and inverted indexing

is represented using a set of unordered discrete acoustic units. The unit can be either word/syllable/phone.

As illustrated in Fig. 3.10, GMM-based soft clustering method is used to model the speech data into a set of K Gaussian clusters. The number of clusters or vocabulary size can be predetermined and is closely related to the number of units (word/syllable/phone) in the language under consideration. Here, each speech segment is clustered into acoustic word with highest posterior probability value. Finally, the speech data is represented as a frequency count or histogram as $[f_1, f_2, \cdots f_K]$, where f_i is the frequency of ith acoustic word in the speech data and K is the vocabulary size. The histogram must be robust enough to mitigate the false-positive and false-negative rate while reducing the time taken for retrieval.

The histograms are then normalized to account the difference in duration of the spoken document. From these normalized histograms, acoustic words having frequency above a threshold value (δ) can be considered as the "significant acoustic words." These significant acoustic words can be used to represent the speech document in the inverted index. Inverted indexing means mapping from contents such as words or numbers (here significant acoustic words) to its corresponding locations in the database. Once the entire database is indexed, it is possible to locate a spoken keyword from the document just by determining the significant acoustic words in them.

3.4 Matching Technique: Dynamic Time Warping

Dynamic Time Warping (DTW) is a widely accepted pattern matching algorithm. DTW is used to find the similarity between two patterns by warping the time axis of one pattern to match with the other. DTW helps to find a region in the test utterance which is highly similar to the query sample. Time warping can handle speaking rate variabilities in two speech patterns. Even though the acoustic similarity between two speech signals can be measured using DTW, it suffers from mismatches due to speaker and/or environmental variations. The template matching can be done either by using spectral-based features (like MFCC, PLP) or by using other symbolic features like posterior features. Many variations are made to the basic DTW algorithm to make it more powerful for audio search applications. Some of the DTW variants are discussed below:

3.4.1 Basic DTW

Consider two patterns **query** : $\mathbf{Q} = (q_1, q_2, \ldots, q_i, \ldots, q_n)$ with n feature frames and **test** : $\mathbf{T} = (t_1, t_2, \ldots, t_j, \ldots, t_m)$ templates with m feature frames. Figure 3.11, illustrates that nonlinear alignment works better than linear alignment. DTW warps/nonlinearly stretches the time axis of one pattern in order to minimize the distance between those patterns as shown in Fig. 3.11b. That is, the DTW finds a warping function $n = w(m)$, in order to map the time axis m of $\mathbf{T}$ with n of $\mathbf{Q}$ [32]. The warping path is derived from the solution of the following optimization problem:

$$D = \min_{w(m)} \left[\sum_{m=1}^{T} d(T(m), Q(w(m))) \right] \tag{3.13}$$

where each d refers to the distance between mth test frame and $w(m)$th/warped query frame, resulting in an $(m \times n)$ distance matrix D. D is the minimum distance measure corresponding to the best path $w(m)$ through the grid of $(m \times n)$ points. From D, the minimum accumulated distance from the starting point (1,1) can be recursively defined as:

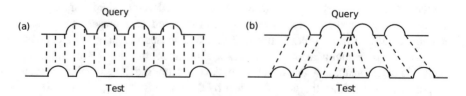

Fig. 3.11 (**a**) Linear alignment. (**b**) Nonlinear alignment

$$D_a(n, m) = d(Q(n), T(m)) + \min[D_a(n - 1, m - 1), D_a(n, m - 1),$$

$$D_a(n - 1, m)] = S(n, m) \tag{3.14}$$

where d is the frame distance at point (m, n). It can also be expressed as *similarity measure*, S. If the patterns have unequal length, it is better to use the longer one on the horizontal axis in order to warp the shorter to the longer. The accumulated distance measure can be normalized with respect to the length of the longer test sequence. Exponential number of warping paths is possible. To reduce the search space, following constraints are used:

1. Monotonicity: $i_{s-1} \le i_s$, $j_{s-1} \le j_s$, where $1 \le i \le n$, $1 \le j \le m$
2. Continuity: $i_s - i_{s-1} \le 1$ and $j_s - j_{s-1} \le 1$
3. Boundary conditions: $i_1 = 1$, $i_k = n$ and $j_1 = 1$, $j_k = m$
4. Warping window $| i_s - j_s | \le | R |$, where $R > 0$ is the window length
5. Slope constraint

Limiting the range $i_{s-1} \le i_s$, $j_{s-1} \le j_s$ is done so as to keep the monotonicity of the warp path (i.e., no temporal backtracking). It ensures that the features are not repeated. Continuity constraint ensures that the alignment path does not jump into "time" index. It guarantees that the alignment does not omit the important features. Boundary conditions constraint the path to include the beginning and ending points. It guarantees that the alignment does not partially consider one of the patterns. Warping window restricts the path wandering too far from the diagonal. It ensures that the alignment will not skip different features and get stuck at similar features. By restricting the warp path within a window of $\pm W$, computational complexity can be reduced. But for good recognition, W must be expanded. The slope constraints control the alignment path not being too steep nor too shallow, thereby prevent the very short parts of the sequences being matched to very long ones. By applying one or more constraints given, different DTW variants can be realized.

Global Constraints

Global constraint helps greatly to reduce the computational complexity by restricting the warping window. This restricts the excessive stretching of query or test or both in order to get an alignment. Global constraints usually considered are Sakoe–Chiba band and Itakura parallelogram as shown in Fig. 3.12. Sakoe–Chiba band fixes the range of time warp and hence known as fixed range DTW algorithm. Let $w_k = (i_k, j_k)$ denote the warping function. As per Sakoe–Chiba band, the relation between i_k and j_k is given by $| i_k - j_k | \le R_s$, where R_s is the speaking rate parameter. Generally, the human speaking rate is not constant, hence using a fixed speaking rate in Sakoe–Chiba band reduces the accuracy. So, in order to overcome this Itakura parallelogram as shown in Fig. 3.12b is used. Itakura suggested a piecewise slope limitation on the signal that can be stretched or compressed. In Itakura parallelogram constraint, the speaking rate R_s is a function of i and j in

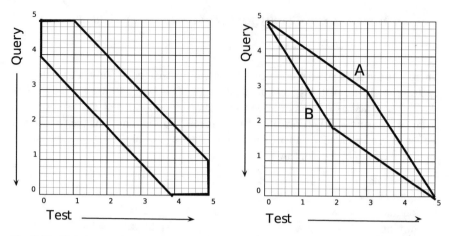

Fig. 3.12 Global constraints considered: (**a**) Sakoe–Chiba band, and (**b**) Itakura parallelogram

the warping path. Therefore, it provides a practical performance improvement in addition to the increase in alignment accuracy without affecting the running time.

Local Constraints

In addition to the global constraints, no single point must be stretched or compressed too much. Local constraints like the type of arc used in the warping path and the weight associated with each branch in the arc are considered. The arc type defines the flow of warping path from the starting frame to the ending frame. Two popular arc types are: Itakura arc and Needleman–Wunsch arc as shown in Fig. 3.13. The similarity matrix S for Itakura arc is defined as:

$$S(i, j) = d(i, j) + \min(w_1.S(i, j - 1), w_2.S(i - 1, j - 1), w_3.S(i - 2, j - 1)) \tag{3.15}$$

The similarity matrix S for Needleman–Wunsch arc is defined as:

$$S(i, j) = d(i, j) + \min(w_1.S(i, j-1), w_2.S(i-1, j-1), w_3.S(i-1, j)) \tag{3.16}$$

The weight assignment can be done symmetrically or asymmetrically. If all branches of the arc have the same weight, it is termed as symmetric weight assignment. In asymmetric weight assignment, weights are assigned so as to give more preference to the diagonal branch than other branches of the arc. For example, let $w1 = 1$, $w2 = 2$, and $w3 = 1$ in Fig. 3.13b in order to give more preference to the diagonal transition.

The problems with basic DTW are its unmanageable computation time and memory requirements. Moreover, it compares two sequences of nearly equal length,

Fig. 3.13 Arc types used in the warping path: (**a**) Itakura, and (**b**) Needleman–Wunsch arc

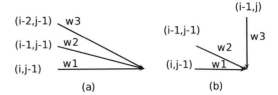

(a)   (b)

but practically the two sequences may have different lengths. Therefore, variants of DTW are suggested by the researchers.

3.4.2 Constrained-Endpoint DTW

As per the boundary constraint, the warping path must include both the starting and ending points of the sequences. Therefore, no freedom is permitted in matching the first and last frames of templates as shown in Fig. 3.12b. This is termed as Constrained-Endpoint DTW (CE-DTW). It is useful for systems whose performance depends on the correctness of endpoint detection [32].

3.4.3 Unconstrained-Endpoint DTW

In most cases, the patterns to be matched will have unreliable end points or might be misaligned. During such situations, the CE-DTW shows poor matching because of the endpoint constraints. In order to overcome these endpoint constraints in CE-DTW, Unconstrained-Endpoint DTW (UE-DTW) was introduced. UE-DTW is a variant of DTW that permits local constraint relaxations up to "x" frames, but only at the start and end locations as shown in Fig. 3.14. By changing the value of relaxation range "x," different variations of UE-DTW can be considered according to the applications. All these methods allow the warping function to start from a point other than $(1, 1)$ and end at a point other than (m, n).

Figure 3.14a shows the unconstrained end point (2:1 slope constraints), where $x/x'/x''/x'''$ frames can be omitted from the start and end points of the patterns. Here, the slope can be varied accordingly to adjust the search space. Figure 3.14b shows the Unconstrained-Endpoint Local Minimum (UELM) method, where the search is limited only to a range around the local optimum path. Compared to CE-DTW, UELM allows more paths at the start so as to avoid the rejection of good paths due to poor alignment but it prunes many paths that CE method would evaluate so as to reduce the search space.

Fig. 3.14 Variants of
UE-DTW: (**a**) UE2-1 (2:1
slope constraint) and (**b**)
UELM
(unconstrained-endpoint local
minimum)

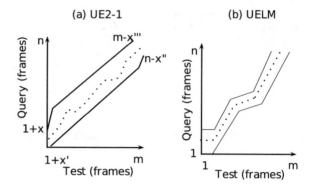

3.4.4 Modified DTW

It helps to find a well-matched path between the query and the test utterance. Two constraints are incorporated here [77]. First constraint disallows simultaneous multiframe path extensions in both query and test utterance as discussed earlier in the continuity constraint. If a DTW path is progressed to index i in the query and j in the test sequence, then it can be extended to $i + n$ and $j + m$, respectively, with a constraint that either $n = 1$ or $m = 1$.

The second constraint is to favor path extensions with similar durations by scaling the similarity scoring. The distance score of path extension is normalized by the number of frames m taken by the test side of the extension. The scaling can be done with the help of an alignment slope factor defined as:

$$\gamma = \max(n, m) \tag{3.17}$$

This alignment slope factor strength can be controlled exponentially by a factor of φ, the duration constraint variable. Therefore, the similarity score of any path extension in the DTW where $m = 1$ can be calculated as:

$$S(q_i \rightarrow q_{i+n}, t_j \rightarrow t_{j+1}) = \gamma^\varphi \sum_{k=1}^{n} D(q_{i+k}, t_{j+1}) \tag{3.18}$$

Similarly, the similarity score for path extension where $n = 1$ is given by:

$$S(q_i \rightarrow q_{i+1}, t_j \rightarrow t_{j+m}) = \frac{\gamma^\varphi}{m} \sum_{k=1}^{m} D(q_{i+1}, t_{j+k}) \tag{3.19}$$

where q, t are query and test utterance vector, respectively.

Thus, it is insured that any final path will have equal contribution from each query frame regardless of the total number of test frames absorbed by the path. Moreover, the duration constraint variable φ can control the path extensions. When $\varphi = 0$, no diagonal alignment constraint is enforced; otherwise, it will force the search to favor perfectly the diagonal path. That means, as φ increases, the value of S also increases and thus such horizontal or vertical path extensions are discarded since it is not the minimum score path. Finally, the DTW finds the minimum scoring path through the similarity matrix.

3.4.5   Segmental DTW

Again, consider $Q = (q_1, q_2, \ldots, q_i, \ldots, q_n)$ and $T = (t_1, t_2, \ldots, t_j, \ldots, x_m)$ as the feature sequence of query and test sequences, respectively. The length of test feature is usually very much greater than that of query ($m \gg n$). The aim is to find a subsequence $X(a'; b')$ such that $1 \leq a' \leq b' \leq m$. Segmental DTW (S-DTW) tries to find out queries or query-like sequence from the test data. S-DTW defines two constraints namely global constraint and local constraint as discussed earlier.

Apart from these global and local constraints, *window shifting* is also a critical parameter. Even though smaller window shifts increase the computational complexity, it gives better performance. So, proper choice of window shift is essential. The step length of the start coordinates of the DTW search can be constrained. If the starting coordinate of the warping path is set, the windowing condition will restrict the shape as well as the end points of the wrap path. Therefore, the warping process with different starting points will result in total of $\lfloor \frac{m-1}{R} \rfloor$ diagonal warping regions with a width of $(2R + 1)$, where R is the window size and m is the length of test utterance. In order to avoid redundancy while computing the warping path, overlapping sliding window strategy is used for the start coordinates as shown in Fig. 3.15. Since the query length is fixed, only the segmentation of test utterance needs to be considered. The start coordinate of each warping segment is given by:

$$(1, (k - 1) \times R + 1)\, ; \; where \; 1 \leq k \leq \lfloor \frac{m - 1}{R} \rfloor \tag{3.20}$$

where each segment shows the warping/alignment between the entire keyword sequence and a portion/subsequence of the test utterance.

Fig. 3.15 Start points of the warping path with $R = 2$

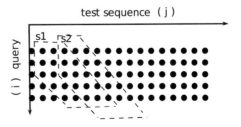

3.4.6 Modified Segmental DTW

Various modified forms of segmental DTW were introduced to reduce the search space and time complexity [46–49]. Segmental Locally Normalized DTW (SLN-DTW) is a variant of S-DTW, where the acoustically similar feature vectors are grouped into a single vector and is considered as a segment [47]. S-DTW is further modified using agglomerative clustering [48]. Memory Efficient Subsequence DTW (MES-DTW) is another variant of S-DTW [49].

Instead of doing S-DTW, computational complexity can be reduced by averaging of features. Modified the sequential DTW matching is later introduced with the average of consecutive features without any overlap [50]. Grouping of features by considering the phone boundaries helps to overcome the information loss that may happen during the feature reduction process.

3.4.7 Non-segmental DTW

Segmentation of spoken audio followed by the DTW is computationally more expensive. Hence, a variant of DTW named as Non-Segmental DTW (NS-DTW) was introduced [76]. The distance measure between query and test sequences in NS-DTW is given by:

$$d(i, j) = -\log \left\{ \frac{q_i}{\|q_i\|} \cdot \frac{t_j}{\|t_j\|} \right\} \tag{3.21}$$

Only the local constraint as shown in Fig. 3.13a is used here. Then, a similarity matrix S of size $m \times n$ is computed. The query term is likely to start from any point in the test data. The similarity matrix will be calculated as follows:

$$S(i, 1) = d(i, 1) \tag{3.22}$$

$$S(i, j) = \min \begin{cases} \frac{d(i,j)+S(i-2,j-1)}{T(i-2,j-1)+w3}, \\ \frac{d(i,j)+S(i-1,j-1)}{T(i-1,j-1)+w2}, \\ \frac{d(i,j)+S(i,j-1)}{T(i,j-1)+w1} \end{cases} \tag{3.23}$$

where i, j indicate the row and column of the similarity matrix, and T is the transition matrix. The elements of T matrix represents the number of transitions required to reach the point (i, j) from the starting point and hence the accumulated score will be normalized with the length of aligned path. T matrix is computed as follows:

$$T(i, j) = \begin{cases} T(i, j-1) + w3 & \text{if } \hat{i} = i - 2 \\ T(i, j-1) + w2 & \text{if } \hat{i} = i - 1 \\ T(i, j-1) + w1 & \text{if } \hat{i} = i \end{cases} \qquad (3.24)$$

where

$$\hat{i} = \underbrace{\arg\min}_{\hat{i} \in i, i-1, i-2} \begin{cases} \frac{d(i,j)+S(i-2,j-1)}{T(i-2,j-1)+w_3}, \\ \frac{d(i,j)+S(i-1,j-1)}{T(i-1,j-1)+w_2}, \\ \frac{d(i,j)+S(i,j-1)}{T(i,j-1)+w_1} \end{cases} \qquad (3.25)$$

A path transition matrix $P(i, j)$ is updated while the similarity matrix is being computed. Path matrix is used to get the start and end locations of the keyword. The *end point* can be located by finding the $\min_j S(n, j)$ where $j = 1, 2, \cdots, m$. Once the end point is located, the corresponding starting locations can be obtained from $P(n, j)$ and hence there is no need to trace back. Path transition matrix is computed as:

$$P(i, 1) = i \quad \text{for } i = 1, 2, 3, \cdots, n$$
$$P(i, j) = P(\hat{i}, j-1) \text{ for } i = 1, 2, 3, \cdots, n \qquad (3.26)$$

If there is more than one query matching location in the reference, NS-DTW helps to find the k-best alignment scoring indices from the similarity matrix. NS-DTW is computationally faster than S-DTW. Time complexity is given by $O(mn) + O(n)$, where O(n) is the time complexity of path trace backing. With $m \gg n$, $O(mn) + O(n) = O(mn)$. Therefore in NS-DTW, time complexity is independent of P matrix. But, P matrix demands higher memory requirements. Other variants of NS-DTW differ in the type of local constraints, values of weights, and frame-level normalization.

A partial matching approach is proposed to retrieve queries that do not appear exactly in the search data [88]. Since the QUESST data consists of three types of queries T1, T2, and T3, where only T1 queries occur exactly same in the search data. So, for T2 and T3 query types, partial matching strategies are employed so that there is no need to run DTW multiple times for each query and test data pairs. For T2 query types, since the query varies either at the beginning or ending, forward/backward partial matching approaches are proposed where the warping path aligns to either the initial or last portion of the warping path. Considering T3 multiword queries, the warping path is broken into two parts and partial matching is done.

3.5   Matching Technique: Minimum Edit Distance

"String distance" is a metric to measure the similarity between two text strings. Minimum Edit Distance (MED) and its variants are the most popular algorithms to find the string distances. MED and its variants are used in audio searching where the audio files are first converted to corresponding text messages using speech recognizers. The MED between two strings (query and test sequences) is the minimum number of editing operations (Insertion, Substitution, and Deletion) needed to transform one string into another.

3.5.1   Conventional MED

Minimum Edit Distance (MED) is defined as the minimum cost of converting one string to the other using the three basic operations [Insertion (I), Substitution (S), and Deletion (D)]. MED is computed by dynamic programming. A distance matrix is calculated with each symbol in one sequence arranged along a row and that of the other sequence along one column. This matrix is termed as edit distance matrix. Each cell in the edit distance matrix can be calculated as a simple function of the surrounding cells. Starting from the beginning of the matrix, it is possible to fill every cell of the matrix. The value in each cell is computed by taking the minimum of the three possible paths. Consider two strings U and V with lengths p and q, respectively. The distance matrix $M(0, \ldots, p)$ $(0, \ldots, q)$ for U and V strings can be calculated as:

$$M(0)(0) = 0 \tag{3.27}$$

$$M(i)(0) = i * I; \quad i = 1, \ldots, p \tag{3.28}$$

$$M(0)(j) = j * D; \quad j = 1, \ldots, q \tag{3.29}$$

$$M(i)(j) = \min\{M(i-1)(j-1) + S(U(i), V(j)), \ M(i-1)(j) + D,$$

$$M(i)(j-1) + I\} \tag{3.30}$$

where S, I, and D are substitution, insertion, and deletion penalties, respectively. The distance matrix is initialized with Eq. (3.27). Top to bottom gradation along a column will increase the insertion penalties whereas row-wise transition will increase the deletion penalties as described by Eqs. (3.28) and (3.29), respectively. For the index (i, j), the minimum penalty either due to substitution, insertion, or deletion is considered as shown in Eq. (3.30). While searching keywords, all strings having an MED lesser than the threshold are declared as hits.

For conventional MED in keyword spotting, substitution penalties are assigned based on the broad acoustic-phonetic classes. The phones substitutions belonging to the same acoustic-phonetic class are considered to be equivalent ($S = 0$) and all other substitutions are declared to be invalid ($S = \infty$). Deletion and insertion penalties are fixed.

3.5.2 Modified MED

In conventional MED measure, substitution penalties are calculated according to heuristic class-based rules. But, it does not actually represent the substitution errors that occur during lattice generation. It is because of the fact that the confusion between different phone models depends on the amount of training data, nature of training data, and richness of the coarticulation captured for the phone. In modified MED measure, the substitution penalties are derived from the phone confusion matrix of the recognizer, since the confusion between different phones can be learned easily from a confusion matrix. It has been reported in [60] that the keyword searching becomes more accurate if we use modified MED measure instead of conventional MED.

If the confusion between two phones (i, j) is more than that between (i, k), it shows that i is more probable to be substituted with j than k. Therefore, the penalty score for the phone pair (i, j) must be fixed lesser than that of (i, k), instead of considering the acoustic classes to which i, j, and k belong to. Finding the substitution penalties from the confusion matrix improves the performance of keyword searching, since it takes care of the real errors made by the recognizer and also smooths out the MED scores.

The simplest way to find the substitution penalty from the confusion matrix is by calculating $1 - C(i, j)$ as the penalty score for the phone pair (i, j), where $C(i, j)$ is the entry in the confusion matrix indicating the number of times the ith phone is substituted by the jth phone. However, it was found that the penalty scores from this simple calculation are very close to each other. So, the substitution penalty for (i, j) phone pair can be calculated as:

$$S(i, j) = \log\{C(i, i)/C(i, j)\}; i \neq j \tag{3.31}$$

Usually, in most cases $C(i, i) \gg C(i, j)$, so the $\log(.)$ is used to keep the penalty scores within limits. Moreover, this calculation gives enough variance among the substitution penalties for various phone pairs than using the simple calculation. It must be noted that $S(i, j) \neq S(j, i)$, which was not the case with conventional MED. Practically, the substitution penalties for only the top M confusing phones may be computed, while for the rest infinite penalty is fixed.

3.6 Summary

All audio search systems contain two basic stages: feature extraction stage and matching stage. A wide variety of studies have taken place with combination of different feature vectors and different matching techniques. In this chapter, basic ideas about such features and matching techniques are discussed in detail. Instead of using these features directly, some representations (like posteriorgrams) derived from primary features are widely used for the keyword searching task. A detailed discussion of various matching techniques like DTW, its variants, and MED is also presented.

Chapter 4
Keyword Spotting Techniques

4.1 Introduction

Keyword spotting (KWS) refers to the spotting and retrieval of predefined keywords
from audio database. Different supervised as well as unsupervised approaches have
been implemented to do keyword spotting. Keyword spotting is considered to be
the first among speech searching. Later, keyword spotting paved the way to Spoken
Term Detection (STD) and Query by Example STD (QbE-STD). In the early days,
researchers have used HMM for KWS, where the speech data is converted into
corresponding text data for text-level matching. But, the latest techniques make use
of MLP and DNN for doing search, so that explicit speech to text conversion is
not necessary. All such techniques for keyword spotting are discussed briefly in this
chapter.

4.2 Techniques for Keyword Spotting

In keyword spotting, the set of keywords used for search will be predefined for
a database. The keywords will be selected so that that words will be occurring
in that database more frequently. Spotting can be done in supervised as well as
in unsupervised manner. A system that uses any labeled information for training
purpose is referred to as supervised. Supervised systems may use either an ASR
system or some pre-trained models. LVCSR-based system has limitations that it
works only for well-resourced languages and has limitations while handling out-of-
vocabulary (OOV) words. For languages where the availability of transcribed data
is limited, it is difficult to do speech to text conversion. Different keyword spotting
techniques are discussed in the remaining sections of this chapter.

Leena Mary, Deekshitha G., *Searching Speech Databases*, SpringerBriefs
in Speech Technology, https://doi.org/10.1007/978-3-319-97761-4_4

4.2.1 Template-Concatenation Model

Technique of word template matching was introduced to the keyword recognition system by Bridle in 1973 [51]. This method was proposed to detect the occurrence of a set of keywords from continuous speech. Incoming speech as well as keywords to be detected are first represented in parametric form. Then, elastic templates of keywords are derived from these parameters. The keyword templates are termed elastic since it can be either expanded/compressed in time in order to account for the speaking rate variations. A keyword is spotted when a segment of the incoming speech is sufficiently similar to the corresponding template. Figure 4.1 illustrates the search using template matching. Every keyword template is matched against every portion of the input and a score is computed. A second stage is required to remove the overlapping putative hits. It is observed that such system suffers from high false alarm rate. Computational complexity for realizing such a system is impractical for searching in large audio database.

An alternative to this is a method to detect keyword as well as non-keyword in the speech [52]. It continually measures the similarity between the input speech and the elastic templates derived from the keywords as explained earlier. In addition to this, the system measures the similarity between the incoming speech and the general Language Model (LM) as shown in Fig. 4.2. Language model is used to evaluate whether the observed speech is a keyword or not. A keyword is said to be detected when the similarity to the keyword template is sufficiently greater than the similarity to the general language model. The general language model consists

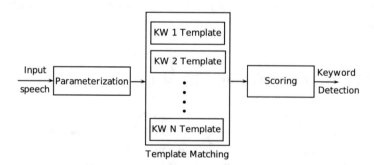

Fig. 4.1 Block diagram of keyword spotting using template matching

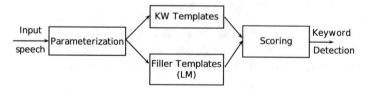

Fig. 4.2 Block diagram of KWS using template-concatenation model

of a set of short segments of speech taken from the training utterance sets. These segments are known as filler templates. Use of filler templates as a general language model is a unique feature of this approach. These short speech segments belonging to the selected subset after statistical clustering are used as filler templates.

4.2.2 Hidden Markov Models

This is a commonly used technique for keyword spotting. Maximum likelihood training of HMM for identified keywords allows the models to assimilate information over many different speakers and word contexts. Hence, it can do better modeling than the template-based systems.

According to [53], an HMM model will be developed for each keyword and a filler model HMM is developed from non-keyword segments of speech signal (fillers). Keywords are represented using linear sub-word acoustic models, to allow large keyword vocabularies, and also to allow robust training of keyword variants that do not appear in training data. Generally, a filler model can be realized either from a fully connected phonetic units or with a LVCSR system where the lexicon excludes the keyword. The latter approach yields a better filler model. These systems make use of Viterbi decoding and hence computationally expensive. In [53], different ways are experimented for developing a better filler model in order to reduce the probability of false keyword detection. Different types of filler models experimented are the following:

- Acoustic word models
- Acoustic sub-word models (Triphone/Monophone models)
- Clustered models
- Vocabulary-independent models

Generally, a normalized word likelihood score is reported for each keyword decoded in the Viterbi network. But, these likelihood scores often exhibit variability in time, making it difficult to separate true keyword hits from false alarms. As a remedial measure, a parallel "back-ground" network of filler models is incorporated. A modified log-likelihood scoring procedure is also suggested to account the sources of variability affecting the keyword likelihood scores. There is a significant improvement in the performance obtained from this likelihood ratio scoring. Moreover, an acoustic class-dependent spectral normalization procedure is suggested to compensate for the linear channel effects [53].

All such fully supervised techniques discussed above are not effective for finding words that are Out-Of-Vocabulary (OOV). Filler- and HMM-based techniques can be modified to search for any number of keywords. But for every new keyword to be searched, keyword model should be trained and the filler model should be retrained. There will be degradation in the performance of those systems, if the test data is acoustically mismatched from the corpora used for training.

4.2.3   LVCSR-Based Techniques

With the help of a Large Vocabulary Continuous Speech Recognition System (LVCSR), the audio files are converted into corresponding text messages. Keywords are also converted to text form. Then, conventional text-level searching techniques are employed to find the occurrence of keywords in the test database. LVCSR approach is common and found to be very accurate for well-resourced languages [54, 55]. But, the main limitation of the system is the need for large amount of transcribed data and inability to handle OOV words.

In an attempt, a speaker-independent continuous speech recognition (CSR) system *(DECIPHER)* was used for the task of keyword spotting [54]. There a keyword is spotted from the one-best output of the Viterbi back-trace. But, this algorithm worked well for keywords with high frequency. The best N outputs and their Viterbi alignments are also used to spot the keywords [55]. Moreover, LVCSR log-likelihood ratio scoring is used to incorporate the language modeling information. This LVCSR log-likelihood approach shows improved detection rate.

4.2.4   Predictive Neural Model

A neural prediction model was proposed for speech recognition purpose [56]. This idea is adapted for the keyword spotting [57]. This idea uses a sequence of Multilayer Perceptrons (MLPs) and has the following advantages:

1. Simple structure
2. Easy to train
3. Training flexibility is possible. Word-/syllable-/phone-level training can be done
4. No need to train non-keyword model
5. Non-keyword is rejected based on the accumulated prediction residual score

Each MLP can predict an actual feature vector based on the preceding and succeeding feature vector frames. Squared Euclidean distance is used to measure the local prediction residual between the actual vector (X_t) and predicted vector $(\hat{X}_t)$. Then, accumulated prediction residual is computed for each keyword. Keyword is detected if the accumulated prediction residual has a value lesser than a threshold value.

4.2.5   Phone Lattice Alignment

Phone lattice-based searching is the most popular approach for keyword spotting. Lattice-based keyword spotting approach is much faster than HMM-based approach, as indexing is done to do searching rapidly. N-best Viterbi recognition is performed on the speech file in order to get the phone lattice representation. The phone lattice compactly encodes multiple observed phone sequence hypotheses for any particular region in the speech. So, this approach provides more flexibility, i.e., even if a keyword is not in the best hypothesis between two nodes, it is still retained in the lattice. Moreover, there will be no concept of vocabulary/OOV words.

For spotting keywords, previously prepared phone lattices are searched. For each node in the lattice, the lattice is transversed backwards to obtain a list of all phone sequences that terminate at the node. Phone sequences that match the target sequence are declared as keyword locations. Since the transversal process is done in text level, it will be very fast.

As illustrated in Sect. 3.3.4, speech utterance is converted to corresponding lattice and by using dynamic programming (DP), keywords are spotted out. This is an efficient solution to keyword spotting as the real-time searching can be done faster. A DP match of the keyword pronunciation against the lattice is also suggested [40] to cut down the computation and false alarms. With the help of cumulative DP matching function, best matching path is constructed, by labeling all keyword phones as either *strong* (exact phone matching) or *weak* (phones may be deleted or substituted). Let the lattice edge be defined as $l = (p_b, p_e, p, s)$, where p is the phone recognized, p_b, p_e are the indices of its beginning and end frames, and s is the score output by the recognizer. If L is the set of lattice edges, then the cumulative DP matching function $C(i, e)$ for the keyword phones $p_1 \ldots p_i$ between some possible keyword start time t and end time e is defined as:

$$\forall t, C(0, t) = 0 \ and \tag{4.1}$$

$$C(i, e) = \max_b \begin{cases} C(i-1, b) + V(b, e, p_i), \\ C(i-1, b) + (1 + e - b)P_s + \max_z V(b, e, z), \\ C(i, b) + P_i, \\ C(i-1, e) + P_d \end{cases}$$

$$\tag{4.2}$$

where P_i, P_d, and P_s are the penalties for phone insertion, deletion, and substitution, respectively, and the function V is defined as:

$$V(b, e, p) = \begin{cases} s, & if (b, e, p, s) \in L \\ -\infty, & otherwise \end{cases} \tag{4.3}$$

Thus, a successful path is constructed for the keyword and is assigned with a score, which is the ratio of keyword path score $C(1, n)$ to the maximum likelihood score of the unknown speech over the same interval.

A very fast and accurate keyword spotting approach is developed in order to overcome the poor miss rate performance of *Phone Lattice Keyword spotting systems* (PLS). Dynamic Match Phone Lattice Search (DMPLS) is built up on lattice-based methods to address the issue of errors with the phone recognizer. This is done by appending the lattice search with dynamic programming sequence matching techniques. Minimum Edit Distance (MED) measure is used to compensate for the insertion, deletion, and substitution errors during lattice searching [58]. MED calculates the minimum cost of transforming one sequence to other, using a combination of insertion, deletion, substitution, and match operations.

Performance of the PLS is poor since the approach needs the entire target sequence to declare a keyword occurrence. Usually, the target phone sequences are erroneous based on the performance of the phone recognizer. Insertion, deletion as well as substitution errors are inherent in a recognizer output. The lattice sequences are accepted/rejected by thresholding on the MED score [58]. PLS is a special case of DMPLS where the threshold of 0 is used. MED calculations used fixed deletion cost and insertion cost ($C_i = 1$). However, substitution cost was decided based on some basic phone substitution rules. For the same letter, consonant phone substitution $C_s = 0$, for vowel and stop substitutions $C_s = 1$, and for all other substitutions $C_s = \infty$. The DMPLS yielded considerable performance benefits over baseline lattice-based PLS systems. The DMPLS can be tuned for application-specific requirements like lower miss or false alarm rates. The use of MED yielded faster speeds without degrading the miss or false alarm rates.

4.2.6   Modified Minimum Edit Distance Measure

As discussed in Sect. 3.5.2, a variation of MED leads to considerable improvement in the performance of keyword searching systems than that of system using conventional MED. In modified minimum edit distance measure, the phone substitution penalties are derived according to the performance of the recognizer. This approach can handle the errors caused by the phone recognizer. Use of Modified Minimum Edit Distance (M-MED) measure results in performance improvement of the system. Figure 4.3 shows the block schematic of keyword spotting using modified MED measure [60].

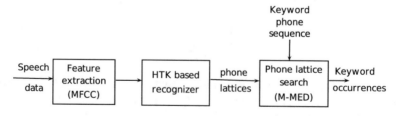

Fig. 4.3 Basic block diagram of keyword spotting using modified MED

4.2.7 Segmental Models

In this approach, a speech recognizer is developed using Segmental Gaussian Mixture Models (SGMM) [59]. This is useful when the availability of resources is extremely limited. In this approach, word-level transcriptions for a very limited amount of data is required (≤ 15 min).

As shown in Fig. 4.4, the speech data is segmented based on the spectral discontinuities. The segments produced are of non-overlapping, variable duration phone-like or syllable-like segments. Then, each sound units are grouped into predefined number of clusters using clustering algorithms. Using these clusters, SGMMs are created. The mean of each mixture term is a time-normalized quadratic trajectory in feature space. An SGMM is a Gaussian mixture model that represents a sound unit of the language. These sound units are termed as "discovered-units" that are phone-like/syllable-like units. Thus, the speech recordings are decoded into transcriptions, in terms of discovered-units. Then, the input speech is converted into SGMM indices (index of the maximum term of the mixture). Till this step, the system is unsupervised. Therefore, the SGMM can be trained on speech recording that are acoustically similar to testing data, thereby eliminating acoustic mismatch.

Then, the transcribed data is used to train a grapheme-to-sound unit converter named "Joint Multigram Model" (JMM). It is used to predict the pronunciation of keywords, in terms of discovered-units modeled by the SGMMs. That means, the JMM act as the pronunciation dictionary. The JMM is used to model the correspondence between the sequence of letters in the transcriptions of train speech and sequence of discovered-units in the segmental model transcriptions of the same speech. So for any given keyword, the text transcription can be converted into sequence of discovered-units, like its pronunciation obtained from a pronunciation dictionary.

Now, the keyword pronunciation can be searched in the discovered-unit transcriptions of test speech. Here, dynamic programming string-searching algorithm (Smith–Waterman) is employed, which minimizes the edit distance between two strings. The resultant edit distance is used as the score for the recording, with lower scores indicating better matches.

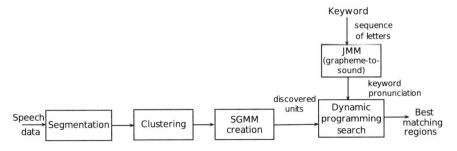

Fig. 4.4 Basic block diagram of keyword spotting using segmental models

4.2.8 *Multilayer Perceptron*

A language-independent approach for audio searching is implemented using a
Multilayer Perceptron (MLP) [26]. MLP is trained using Wall Street Journal corpus
[14]. Using that MLP, keyword spotting of English and Hindi languages is achieved
with reasonable accuracy. Phonetic posteriorgrams are generated from MLP for each
input frame and a modified DTW algorithm is used for searching. Posteriorgram-
based audio search framework trained using English language is used for Hindi
without any modification, thus attaining a language-independent search approach.
The testing and training procedures are illustrated in Fig. 4.5.

MFCC features (39-dimensional) are extracted from each speech frame. MLP
is fed with MFCC features of nine successive frames as input and phonetic label
of middle frame as output. Structure of MLP is 351 input nodes, 1 hidden layer
with 1000 nodes, and output layer with 40 nodes [26]. With the help of MLP, a
given utterance is converted into phonetic posteriorgrams which is the representation
of sequence of probability vectors over time. A posteriorgram is a 2-dimensional
representation with time on X-axis and phonetic class probability on Y-axis.

A modified DTW algorithm described in Sect. 3.4.4 is used to find the best
possible path corresponding to the query in each of the audio database utterances
with less computational complexity. Distance measures used are dot product and
KL divergence. Experiments showed that KL divergence performs better than dot
product. The accuracy increases as the number of hidden neurons increases at the
expense of more computation time.

Fig. 4.5 Basic block diagram
of language-independent
audio searching using MLP
and modified DTW

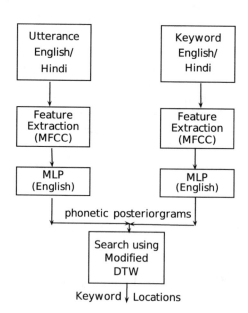

4.2.9 Deep Neural Networks

Deep Neural Networks (DNN)-based KWS is proposed to realize a system with small memory footprint, low computational cost, and high precision [72]. DNN is trained to directly predict the keywords or its sub-word units followed by a posterior handling unit producing a final confidence score as shown in Fig.4.6. Unlike HMM systems, this approach does not require a decoding unit, so that it will be a simpler implementation with less runtime computation and smaller memory footprint. Even though trained using small amount of data, the system outperforms HMM system for clean as well as noisy speech.

To reduce computational complexity, voice-activity detection is used to run the algorithm only on voiced regions. Perceptual Linear Prediction (PLP) features along with their deltas and double-deltas are used as input. For DNN-based KWS, the output labels represent entire words or sub-word units. Results with full word labels outperform the corresponding sub-word units and are computationally less expensive [72]. Three hidden layers having 128 nodes per layer with Rectified Linear Unit (ReLU) activation function and the last layer with *softmax* activation function are used [72].

The DNN posteriors are combined into keyword/key-phrase confidence scores for better results. For computing the confidence score, the posteriors are smoothed over a fixed time window of size $w_{smooth} = 30$ frames. Smoothed posterior p_{ij} of ith label and jth frame is calculated by:

$$pt'_{ij} = \frac{1}{j - h_{smooth} + 1} \Sigma^{j}_{k=h_{smooth}} p_{ik} \tag{4.4}$$

where $h_{smooth} = \max(1, j - w_{smooth} + 1)$ is the index of the first frame within the smoothing window. Then, the confidence score at the jth frame is computed within a sliding window of size $w_{max}=100$ as follows:

$$confidence = n - 1 \sqrt{\Pi^{n-1}_{i=1} \max_{h_{max} \leq k \leq j} pt'_{ik}} \tag{4.5}$$

where $h_{max} = \max(1, j - w_{max} + 1)$ is the index of the first frame within the sliding window. Finally, a decision will be made if the confidence score exceeds some threshold value.

Convolutional Neural Networks (CNN) is the latest deep learning technique which is used for the speech search task [3, 25, 73, 74]. Studies have proved that

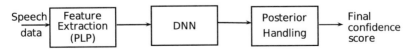

Fig. 4.6 Block diagram of keyword spotting using DNN

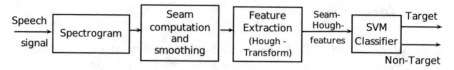

Fig. 4.7 Block diagram of keyword spotting using seam patterns

CNN based systems can outperform the DNN based systems and can develop models with smaller size [73]. CNN based systems can perform spoken term detection from noisy as well as low resourced languages. Bottleneck feature representation helps to reduce the dimensionality and it gives noise robustness [74].

4.2.10 Spectrographic Seam Patterns

The basic block of keyword spotting using spectrographic seam pattern is shown in Fig. 4.7. Seam–Hough feature vectors are extracted from the speech signal as discussed in Sect. 3.3.1. Since spectrograms are more tolerant to fine-grained variations in texture, the seams can be smoothed using a filter. Features for each word are then derived using a variant of Hough transform. A discriminative classifier is built for these features using SVM to distinguish between target and non-target words as shown in Fig. 4.7.

Seam–Hough features extracted from different instances of keyword are used as positive examples for training the SVM classifier. Features from speech recordings that do not contain the target word are taken as negative samples. For word spotting, a sliding window is used for feature extraction and classification. If adjacent windows are classified as a keyword, then these windows are merged into a single unit [38]. The classifier performs well as the number of seams is increased [38].

4.2.11 Spectro-Temporal Patch Features

As discussed in Sect. 3.3.2, the spectro-temporal patch features are used for keyword spotting [39]. A patch dictionary consisting of the patches is extracted, and their center location is constructed. Using this dictionary, *patch dictionary response* $\{R_k\}_{k=1}^K$ is computed from the spectrograms $S(f, t)$ of duration T. Each patch in the dictionary is placed at the location $(f_k, rt_k * T)$, and *norm* is computed between the patch and the corresponding portion of the spectrogram as:

$$R_k = \sum_f \sum_t \| P_k(f - f_k, t - rt_k * T) - S(f, t) \|^2 \tag{4.6}$$

where f_k is the patch center frequency and rt_k is the relative time location. Equation (4.6) describes an operation similar to convolving the spectrum of target keyword with that of the input spectrum. Thus, if there are K patches in the dictionary, then the dimension of the spectro-temporal response $R(k)$ will also be K, irrespective of the length T of the spectrogram $S(f, t)$. While these patches being *selective* (i.e., the system will respond selectively if the target word occurs), it must be *invariant* too. So, a pooling operation is done to account for the patch frequency and temporal shifts. In [39], *MAX/MIN* operation is suggested to overcome the typical patch variations seen in images. A local frequency shift range ΔF_{range} and a local temporal shift range ΔT_{range} are defined, so that the pooled patch response $\hat{R}_k$ is defined as:

$$\hat{R}_k = \min_{\|f_c - f_k\| \le \Delta F_{range}, \|t_c - rt_k * T\| \le \Delta T_{range}} R_k(f_c, t_c) \tag{4.7}$$

Thus, these time-invariant, K-dimensional, pooled patch dictionary responses $\hat{R}_k$ can be used to learn a decision function to separate between positive keyword patch responses from negative non-keyword patch responses. An SVM classifier is used to distinguish between positive patch responses and negative responses [39].

4.3 Utterance Verification

Different methodologies have been adapted in different languages to spot the keywords from spoken database. Most of the systems perform well for very long keywords, but their performance will degrade for shorter keywords. The basic problem with keyword spotting systems is the higher false alarm rate, especially for shorter keywords. Hence, a second stage is preferred to verify the utterance identified by the first stage. An isolated keyword verification system helps to reduce the false alarm.

A good keyword spotting system also implies a good non-keyword rejection system. In tonal languages like Mandarin (also syllabic language), words can be represented using a single syllable. Short keywords are very difficult to verify since they contain only less information and they may be confused more easily with other non-keyword speech. Most of the systems make use of log-likelihood ratio for keyword verification. Verification system can be used at different levels like syllable level, word level, etc.

Figure 4.8 shows the basic block diagram of a keyword spotting system. It is a two-stage system in which the first stage detects the keywords roughly and the second stage verifies whether the short-listed regions are keywords itself. The keyword verification stage reduces the false alarm rates and helps to come up with the actual keyword locations.

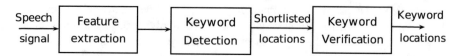

Fig. 4.8 Block diagram of utterance/keyword verification system

4.3.1   Confidence Measures

Some verification systems make use of different confidence measures such as verification function. In [62], four such confidence measures are investigated. Keyword is recognized using LPC feature-based HMM syllable models. Moreover for each syllable, an anti-syllable model is also trained. It is developed by grouping the highly confusing syllables. Also, a general acoustic filler model was developed on non-keyword speech data and a background/silence model is trained on non-speech segments of input signal.

In the first stage or recognition stage, Viterbi algorithm is employed to find the most likely keyword W_K, where

$$W_k = \arg \max_j L(O/W_j) \qquad (4.8)$$

and $L(O/W_j)$ is the likelihood of the observation sequence O for a given word W_j. W_k is the combination of syllable sequence and can be written as:

$$W_k = s_1^{(k)} s_2^{(k)} \cdots s_N^{(k)} \qquad (4.9)$$

Then in the second stage, utterance-based confidence score is calculated by combining syllable-level likelihood scores. For each syllable, the likelihood ratio (LR) is defined as:

$$LR_n = \frac{P(O|H_0)}{P(O|H_1)} = \frac{P(O|\lambda_n^c)}{P(O|\lambda_n^a)} \qquad (4.10)$$

where O is the observation sequence, H_0 is the null hypothesis that syllable is present in the speech segment, and H_1 is the other hypothesis that syllable is not present in the speech segment. λ_n^c and λ_n^a denote the corresponding syllable and anti-syllable model for the syllable n. By taking the logarithm of (4.10) and normalizing it by the duration of the speech segment (l_n),

$$LLR_n = \log\ P(O|\lambda_n^c) - \log\ P(o|\lambda_n^a)/l_n \qquad (4.11)$$

LLR shows the exact recognition score offsetted with a score computed with the anti-syllable model. Confidence measure of a word is defined as a function of likelihood ratios of syllables in that word. A keyword will be accepted as valid, only if the confidence measure exceeds a threshold value.

$$CM = f(LLR_1, LLR_2, \cdots, LLR_n) \tag{4.12}$$

The first confidence measure CM_1 is based on frame duration normalization and is defined as :

$$CM_1 = \frac{1}{L} \Sigma_n (l_n * LLR_n) \tag{4.13}$$

The second one CM_2 is based on syllable segment-based normalization. It is a simple average of log likelihood of all the syllables and is given by:

$$CM_2 = \frac{1}{N} \Sigma_n LLR_n \tag{4.14}$$

The third one CM_3 focuses on less confident syllables rather than averaging all. In-order to find the less confident syllables, the log-likelihood ratio is normalized assuming a Gaussian distribution to each syllable. The normalized log likelihood is defined as:

$$LLR^* = \frac{LLR_n - \mu(LLR_{c(n)})}{\sigma(LLR_{c(n)})} \tag{4.15}$$

where $\mu(LLR_{c(n)})$ and $\sigma^2(LLR_{c(n)})$ are the mean and variance for syllable class of n. Thus, syllables whose likelihood ratios are less than their means are shortlisted and CM_3 is defined as:

$$CM_3 = \frac{1}{N} \Sigma_n \begin{cases} LLR_n^* & \text{if } LLR_n^* < 0 \\ 0 & \text{else} \end{cases} \tag{4.16}$$

The fourth confidence measure makes use of the sigmoid function as a loss function for training with the minimum error rate criteria.

$$CM_4 = \frac{1}{N} \Sigma_n \frac{1}{1 + \exp(-\alpha(LLR_n - \beta))} \tag{4.17}$$

For each confidence measure, a threshold is set. If the value is below the threshold, that word will be discarded. Experiment shows that CM_3 gives the best performance [62].

4.3.2 Hybrid Neural Network/HMM Approach

This is also a two-stage system: primary keyword spotting stage followed by a verification stage. Usually, hybrid neural network/HMM methods feed the neural networks directly, with the features extracted from input speech. But in [63], the posterior likelihood generated by a set of HMM is fed to the networks. This approach is motivated by the successful application of HMM in different studies and by the strong discrimination ability of neural networks. The system architecture is shown below in Fig. 4.9

As shown in Fig. 4.9, for each keyword, a classifier is built, including a set of HMM modules and neural network. The classifier classifies each putative hit as either a true hit or a false alarm. MFCC features are used in the primary stage. Let O be the acoustic observation sequence of a speech utterance segmented by the word spotter for the keyword w. The observation sequence O is first fed to four modules named "K_w," "A_w," "Fil," and "Dur." "K_w" means a whole-word HMM for keyword w. "A_w" is an "Anti-keyword HMM" modeling miss-recognition. "Fil" is the filler model to model non-keyword speech, which is a set of HMMs derived through clustering of context-independent phones together with a silence model. "Dur" is a function to get the word duration of the observation sequence O.

The anti-keyword model is trained with the help of observation vectors that are recognized wrongly as keyword by the primary spotter. The higher false alarm rate of the spotting system helps to collect the sufficient samples for training anti-keyword model. Then, the likelihoods $L(O \mid K_w)$, $L(O \mid A_w)$, and $L(O \mid Fil)$ together with duration $Dur(O)$ are given as input to the neural network.

The output of the network P_t and P_f corresponds to true hit and false alarm, respectively. Ideally, $P_t = 1$ and $P_f = 0$ correspond to a true hit, and just reversed for a false alarm. If $(P_t - P_f) > \theta$, where θ is the threshold, the hit will be accepted as a true hit. Experiments show that this approach outperforms the baseline systems. It is noted that Dur plays an important role in utterance verification of short words [63].

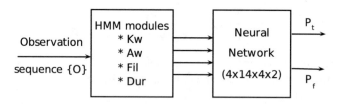

Fig. 4.9 System architecture of hybrid neural network/HMM classifier

4.3.3 Cohort Word-Level Verification

The Cohort Word-level Verification (CWV) helps to improve the performance of a verification system by adding linguistic and word-level information in the selection of non-keyword models. Cohort keywords are words that have similar pronunciation to the target keywords (e.g., verifying whether DINER is DONER or so). Use of similarly pronounced words as non-keywords helps in the selection of non-keywords at word level. Since it is a word-level process, linguistic and word-level information is incorporated implicitly into the selection process.

The cohort words in a language can be selected using a string distance algorithm named Minimum Edit Distance (MED). In CWV method, a confidence score is calculated based on the difference between the target word and similarly pronounced words. Minimum edit algorithm is a set of operation done to transform a sequence W to a sequence C, with the expense of minimum cost.

$$S = s_1, s_2, \cdots, s_k, \cdots, s_K = MinimumEdit(W, C) \tag{4.18}$$

where $s_k \in$ (insert, delete, substitute, match). Each operation (op) is associated with a cost function $\psi(op)$. The overall minimum transformation cost or minimum edit distance is calculated as:

$$d(W, C) = \Sigma \psi(s_k) \tag{4.19}$$

Any words that have a MED score within a predetermined range are used as cohort words. For a given keyword and a corresponding set of cohort words $Q = q_1, q_2, \cdots, q_B$, verification is done similar to log-likelihood ratio method. Confidence score is calculated using LLR method. Some variants of LLR are also addressed like mean cohort word LLR and maximum cohort word LLR as given below:

$$LLR = \log P(O/\lambda_{kw}) - \frac{1}{B} \Sigma_{i=1}^{B} \log P(O/\lambda_{q_i}) \tag{4.20}$$

$$LLR = \log P(O/\lambda_{kw}) - \max_{i=1}^{B} \log P(O/\lambda_{q_i}) \tag{4.21}$$

where O is the observation sequence, λ_{kw} is the keyword model, and B is the number of cohort words.

In [64], HMM models are developed for keywords and cohort words, using 39-dimensional MFCC features. Evaluations are done on acoustically similar as well as dissimilar data sets. Four types of evaluation procedures are done, namely: (1) baseline background model-based system, (2) mean log-likelihood of cohort words, (3) maximum log-likelihood of cohort words, and (4) rank of target keyword within cohort word scores. It has been observed that the verification performance using acoustically dissimilar words outperforms to words that are acoustically similar.

All cohort word verification methods especially ranking method outperform the baseline systems. Experiments are done to evaluate joint verifiers that combine the baseline verifier and cohort word verifiers. Joint verifiers have significantly better performance than the baseline verifier.

4.4   Summary

Various approaches for keyword verification are discussed in this chapter. These approaches aim to reduce the false alarm rates of the keyword spotting systems especially for shorter keywords. Verification systems act as a second stage of keyword spotting systems. Most of the verification tasks make use of log-likelihood ratios or its variants as confidence score. In the literature, there are approaches which use phonetic and prosodic information for utterance verification, mainly for tonal languages.

Chapter 5
Spoken Term Detection Techniques

5.1 Introduction

As discussed, keyword spotting technique refers to the process of searching and retrieving a closed set of keywords. The technology has now improved to search and retrieve any unspecified spoken word from an audio database with reasonable accuracy. This is termed as Spoken Term Detection (STD). STD can be broadly classified into two: (1) Text-based STD systems or simply STD and (2) Query by Example STD (QbE-STD). In normal STD, query as well as database is converted to corresponding text/symbol for searching purpose. Since this technique needs speech to text mapping, some form of recognition systems is used. So, it is a challenging task for under-resourced languages. In QbE-STD, the query is processed in audio format. Speech to text conversion is not used for searching the database. There is no limitation on the query word length and its frequency. Based on the way of realization, STD systems are categorized as supervised and unsupervised.

5.2 Spoken Term Detection Using Supervised Learning

5.2.1 Phonetic Posteriorgram Templates

Figure 5.1 illustrates the realization of QbE-STD using phonetic posteriorgrams [77]. A phonetic recognition system developed at the Brno University of Technology (BUT) [61] is used to generate phonetic posteriorgrams for both query and test utterances. Then, the similarity between these two posteriorgrams is computed.

Let Q represent the query posteriorgram with n frames of speech segment and X refer to the posteriorgram of test utterance containing m frames.

$$Q = q_1, \ldots, q_n \tag{5.1}$$

© The Author(s), under exclusive licence to Springer Nature Switzerland AG 2019
Leena Mary, Deekshitha G., *Searching Speech Databases*, SpringerBriefs
in Speech Technology, https://doi.org/10.1007/978-3-319-97761-4_5

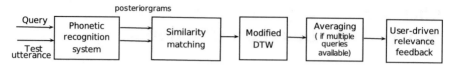

Fig. 5.1 Block diagram of QbE STD using phonetic posteriorgram templates

$$X = x_1, \ldots, x_m \tag{5.2}$$

Then, distance between Q and X is defined as:

$$D(q, x) = -\log(q.x) \tag{5.3}$$

Values of D near to zero show strong similarity between the test and query sequence, while large value shows dissimilarity. This measure will fail if many of the values of q and x are zero, resulting in $q.x = 0$, which leads to $D = \infty$. To compensate for this, the posteriorgrams of both test and query samples are smoothed as follows:

$$q' = (1 - \lambda)q + \lambda u \tag{5.4}$$

where u represents a uniform probability distribution vector and $\lambda > 0$ assures nonzero probability for all phonetic posteriors in q'. Thus, the query and test posteriorgrams are compared and an $n \times m$ similarity matrix is obtained by computing the similarity between the posterior distribution of all n query frames against that of all m test frames. Modified DTW algorithm is then used to find a well-matched path between the query and the test utterance. Thus, DTW search finds the minimum scoring path through the similarity matrix.

There are two approaches to effectively use the available multiple examples of queries. First approach is to combine all the templates into a single template so that it can reflect the characteristics of those multiple queries. Second approach is rather simple but computationally expensive method where all the queries available are used to generate scores. These scores are combined together by score fusion/averaging.

5.2.2 Segment-Based Bag of Acoustic Words

Figure 5.2 shows the method for QbE-STD using segment-based Bag of Acoustic Words (BoAW) technique. First the features are extracted. Using BoAW, significant acoustic words are identified and will check the index to find the keyword locations in the database. A histogram score is calculated for each segment in the database. It is the number of times that segment is retrieved by the significant acoustic words of the query segment. Higher the histogram score, higher the probability of the segment

Fig. 5.2 Block diagram of
QbE STD using Bag of
Acoustic Words (BoAW)

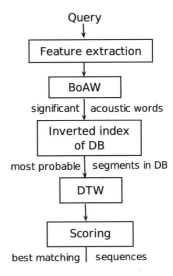

being a correct match. As the BoAW approach ignores the sequence information in speech while indexing, the system performance will decrease due to higher false acceptance. Therefore, Dynamic Time Warping (DTW) is used to restore the sequencing information in the retrieved data segments. DTW is performed between the Gaussian posteriorgrams of query and that of the most probable database segments. The DTW distance function used is:

$$D(p, q) = -\log(p.q) \tag{5.5}$$

where p and q are two Gaussian posterior vectors. Thus, the ranking of the database segments is performed not only using the BoAW histogram score but also using the DTW score. The merged score S_{M_i} of the database segment S_i is computed as:

$$S_{M_i} = \alpha.S_{DTW_i} + \frac{\beta}{S_{Hist_i}} \tag{5.6}$$

where α, β are the scaling factors and S_{DTW_i}, S_{Hist_i} are the DTW and histogram scores of the segment S_i, respectively. The keyword locations are supposed to have lower values of merged score. Thus, the database segments are ranked in ascending order of their merged score.

Figure 5.3 illustrates an approach for QbE STD with the help of partial matching and search space reduction [88]. As shown in Fig. 5.3, the audio data as well as the query is first converted to corresponding posteriorgrams with the help of *BUT* phoneme recognizer [61]. These posteriorgrams are grouped by averaging them out according to the phone boundary. The phone boundary is determined with the help of *Spectral Transition Measure (STM)*. STM is preferred over the phoneme decoders for phone segmentation, because it does not require training; moreover, the phone

Fig. 5.3 Block diagram of
QbE STD using BUT
phoneme recognizer and
BoAW

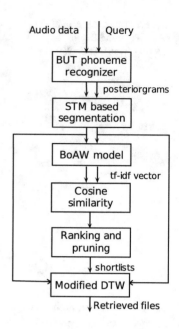

decoder is a language-dependent system, which is trained for a particular language.
Then, BoAW approach is used further to reduce the search space. It is a two-stage
process. In the first stage, cosine similarity score is computed between the BoAW *(tf-
idf)* vector of audio data and query, respectively. The term frequency *(tf)* represents
the total number of times the term t is present in the document. The inverse
document frequency *(idf)* for each term is given by $idf(t) = \log\left(\frac{N}{df_t}\right)$, where
df_t is the number of documents that contains term t and N refers to the total number
of documents. Thus, a set of test utterances are shortlisted and DTW is performed
only on these selected utterances for doing QbE-STD. Partial matching strategies
are adopted in modified DTW algorithm in order to address the realistic scenarios,
where the query does not appear exactly in the search data. The queries might have
different prefix, suffix, or word order, so different strategies are employed to avoid
DTW multiple number of times for each query and test utterance pair.

5.2.3   Phonetic Decoding

A phonetic decoder and Dynamic Match Lattice Spotting (DMLS) technique is
useful to quickly find the search terms [66]. The system has mainly two stages as
shown in Fig. 5.4. In the *indexing* stage, the database is indexed and arranged after
phonetic decoding. Indexing is used to construct the database, so that it can provide
fast and robust subsequent search. In the *search* stage, the keyword is converted
to corresponding phonetic representation and DMLS is used to locate the similar

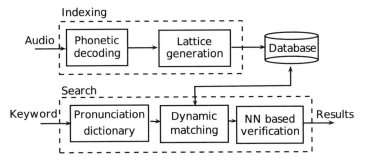

Fig. 5.4 Block diagram of STD using phonetic searching approach

locations [67]. If the keyword is an OOV word, letter-to-sound rules are used to get the corresponding phonetic pronunciations. If the phones are mapped to their corresponding classes (like vowel, nasal, etc.), these hyper-sequences are used for dynamic matching. This helps to reduce the search space and time. Minimum Edit Distance (MED) is used to locate a matching sequence from the database. The score of occurrence of the keyword is estimated by linearly fusing the MED score with the acoustic log likelihood ratio score, *ALLR(P)*, as follows:

$$Score(P, Y) = MED(X, Y) - \alpha.ALLR(P) \qquad (5.7)$$

where X is the indexed phone sequence, Y is the target sequence, P_x represents the individual occurrences of X such that $P \in P_x$ is an indexed phone sequence, and α is an empirically tuned constant. Fusion of ALLR helps to differentiate between occurrences having the same MED score and to promote the occurrence with more acoustic probability. Since the MED score is not comparable for occurrences having different phone lengths, an NN-based verification stage is used to produce a final detection confidence score for each putative term occurrence.

Since the search in phone lattice-based STD is slower, [68] proposed a faster method with smaller index size. Even though the proposed system performance is inferior to other approaches using phone lattices, it can do fast searching. In order to achieve the goal of small index and faster search speed, only the one best phoneme sequence from the phone lattice is considered. Then, a *probabilistic pronunciation model* is proposed to compensate the errors during phoneme recognition [68].

STD using fast phonetic decoding is proposed using DMLS technique [66, 69]. But here, the goal is achieved through monophone open-loop decoding along with fast hierarchical phone lattice search. Use of monophone open-loop decoding helps to achieve the goal of rapid searching. As discussed above, the preliminary search is done using the hyper-sequence database (HSDB), to reduce the search speed and search space. Then, the second MED-based search is done using the actual sequence database (SDB) corresponding to the short-listed locations of preliminary search.

A method based on the joint alignment of phone lattices generated from the query as well as search utterance has shown that it is better to use more than 1-

best phone strings for the utterance or the query [70]. A Weighted Finite State Transducer (WFST)-based searching and indexing system is proposed, where it allows to use the lattice representation of the audio sample directly as query to do the searching [44]. FST are often used for the phonological and morphological analysis in Natural Language Processing (NLP) and applications. By comparing different phone representations derived from word and hybrid (word and sub-word units) systems, it is shown that the hybrid system performs well for OOV words. Similarly, the combination of phone and word confusion networks is useful for open-vocabulary spoken utterance retrieval [43]. Use of confusion network helps to have more compact index table so as to do more robust keyword matching compared to typical lattice-based techniques.

5.3   Spoken Term Detection Using Unsupervised Learning

5.3.1   Gaussian Posteriorgrams

Feasibility of unsupervised learning framework for the keyword spotting task is demonstrated using Gaussian posteriorgrams [79, 83]. Here, a Gaussian Mixture Model is trained as in Fig. 5.5. Each speech frame is labeled into corresponding Gaussian posteriorgrams using GMM. If the training data is noisy or contains some non-speech artifacts, the training of GMM will be difficult due to their large variance. Then, the modeled posteriorgrams cannot well discriminate between phonetic units. So, the training data is first classified as speech or non-speech segments. Speech segments are then used to train GMM.

DTW variants are used to compare the Gaussian posteriorgrams of keyword samples with that of the unseen test utterances. Then, the results are ranked based on the distortion scores.

QbE-STD can be addressed using segmental DTW of Gaussian posteriorgrams as shown in Fig. 5.6. A GMM is created and Gaussian posteriorgrams of query and test data are computed using the GMM [83]. These two posteriorgrams are aligned using

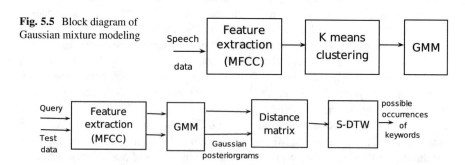

Fig. 5.5 Block diagram of Gaussian mixture modeling

Fig. 5.6 Block diagram of QbE STD using Gaussian posteriorgrams

S-DTW algorithm. A distance matrix is constructed using negative log magnitude metric. Then, similarity matrix is derived from this distance matrix using S-DTW and possible occurrences of queries are located from test database.

S-DTW with two constraints can be used for QbE-STD [79]. First one is the adjustment of the window condition ($| i_k - j_k | \le R_s$), which prevents the warping path from going too far ahead or behind in either posteriorgrams. The second constraint is to apply different starting coordinates to the warping process, so as to divide the difference matrix into different diagonal regions. Overlapped sliding windowing helps to avoid computational redundancies. Finally, a region from test utterance with minimum distortion score is chosen for a particular query.

In [76], QbE-STD is addressed using Gaussian posteriorgrams obtained from several spectral as well as temporal speech features. Spectral/temporal features extracted from the speech signal are used to get the corresponding Gaussian posteriorgrams. Query as well as search data is converted to corresponding posteriorgrams. Searching and retrieval is done using variants of DTW like NS-DTW and fast NS-DTW (3.4.7) as shown in Fig. 5.7.

Fourier–Bessel Cepstral Coefficients (details given in Sect. 3.2.2) is used to train a GMM in order to get Gaussian posteriorgram representation [33]. Figure 5.8 shows how an unsupervised technique is used for QbE-STD. During the training phase, GM components are initialized using k-means clustering method. The clustering will be initialized by finding the mean of the entire training data. The variance is diagonal and is initialized as the identity matrix. The Expectation Maximization is used. For getting the GPs, a row probability vector is generated for each speech frame, and zero mean normalization is also employed. Finally, for finding the possible locations of keyword in the test utterances, S-DTW is employed.

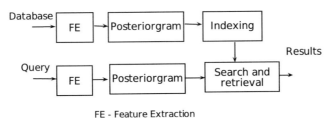

FE - Feature Extraction

Fig. 5.7 Block diagram of QbE STD using FDLP and NS-DTW

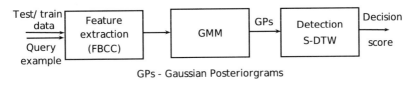

GPs - Gaussian Posteriorgrams

Fig. 5.8 Block diagram of QbE STD using Bessel features

5.3.2 ASM Posteriorgrams

Acoustic Segment Modeling (ASM) technique uses posteriorgrams for performing unsupervised QbE-STD [82]. Both the spoken query and search data is converted to corresponding ASM posteriorgrams. Then, the possible query occurrences are located using ASM posteriorgrams-based template matching technique. Segmental DTW is used in [82] to match ASM posteriorgrams.

Even though posteriogram based matching shows good accuracy, it is computationally complex and takes more time for retrieval when frame level posteriorgrams are considered. Subword or phoneme level posteriorgram matching reduces the computation on the expense of loss of frame information. Obara et al. [80] suggests two methods to improve the speed of QbE using posteriograms of DNN. The methods are, either by transforming the posteriorgram to a bit matrix or by using the sparse vector method to replace the less probable elements with 0. Ram et al. [81] exhibits a faster good performing system that depend on sparse subspace modeling to reduce the dimensionality.

5.3.3 Group Delay-Based Segmentation

In this approach, the speech data is segmented into smaller acoustic units using *Group Delay* (GD) functions [76]. The segmentation boundaries can be varied from phonemes to words by using a parameter named *window scale factor* in the GD-based Segmentation (GDS) algorithm. This method of segmentation has the following advantages:

- No prior knowledge required
- Fast algorithm
- Seamless operations on OOV words
- Language- and text-independent approach

Keyword detection done with the help of group delay-based segmentation is as shown in Fig. 5.9 [85]. Boundaries predicted by the segmentation algorithm may contain misalignments. So, the boundaries will be expanded by a few frames on either side. To find the keyword locations, each segment is passed through DTW-based template matching algorithm. Keywords are located if the distance measure

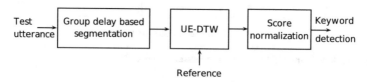

Fig. 5.9 Block diagram of keyword spotting using group delay-based segmentation

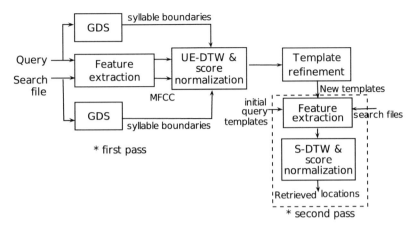

Fig. 5.10 Block diagram of QbE-STD proposed for zero resource languages

lies within the threshold, else the segment will be rejected. In [85], UE-DTW is used to remove the endpoint constraints. Score normalization is done to reduce the false alarm rate.

A fast QbE-STD is also proposed using the Group Delay-based segmentation algorithm [86]. As shown in Fig. 5.10, the proposed system uses a two-pass strategy to spot the keywords with MFCC as feature vectors. In the first pass, some query matches from the search file are identified and these are used as new query templates for the second pass along with the initial query templates. Three manually chosen keyword templates are used as input in the first pass. As in Fig. 5.10, GDS is used to get the syllable-level segments of the search file as well as the query files. During the first pass, the query templates are searched at each syllable boundaries of the search file. UE-DTW is used to find new matching templates. Thus, for the three input query templates, three new templates are obtained. So, all these six templates are used in the second pass to do S-DTW on the search file. By using geometric mean of all the DTW scores, a good match of the keyword is located in the search file. Since the queries used are of different lengths, the DTW score is computed and normalized according to the number of syllables in the query.

5.3.4   Morphological Image Processing Techniques

Morphological image processing techniques have been proposed to spot the keywords from continuous speech signal [87]. Figure 5.11 shows the block diagram for doing STD using morphological image processing techniques. The speech data is converted to corresponding phone posteriorgrams using a hybrid HMM–ANN phone recognizer. ANN-based posteriors give the smooth representation of the speech signal. Contextual MFCC is used as the feature vector by padding 4

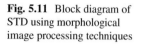

Fig. 5.11 Block diagram of STD using morphological image processing techniques

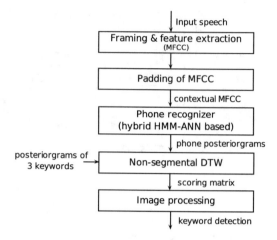

left and right neighboring frames all together. A variant of DTW is applied to the posteriorgrams of keyword and test sequence. DTW distance matrix is then represented as a gray-scale image and is processed further to extract the diagonal streak from it showing the keyword presence. The set of operations performed on the DTW image include segmentation, inversion, dilation, skeletonization, and erosion. This is followed by a connected component analysis.

Dynamic threshold-based segmentation is done to convert the gray-scale accumulation matrix to binary image. Inversion helps to get the desired objects from background to foreground. Dilation with some structuring element helps to repair the broken connections occurred while doing segmentation. The dilated image is then skeletonized to one pixel thick representation in the binary image. Erosion helps to subtract the false horizontal/vertical branches from the skeleton image. Finally, the connected component analysis helps to identify the presence of keyword in the continuous speech signal. The keyword is identified if the foreground object has a portion of it in the last row of the image.

5.4 Summary

In this chapter, we have discussed various techniques for STD. An STD technique is referred to as QbE-STD if the query is processed in the audio format, without explicit conversion to text form. STD techniques mentioned here are categorized into supervised and unsupervised. Supervised techniques mostly use some form of token recognition systems as front end and the output is represented in terms of phonetic posteriorgrams, BoAW, or phone lattices. Unsupervised STD techniques represent tokens in terms of Gaussian/ASM posteriorgrams. All these supervised as well as unsupervised techniques mostly employ variants of DTW for verifying the keyword occurrences at a later stage. Of late, different types of image processing techniques are also proposed for the verification stage of STD.

References

1. Wang LC (2003) An industrial-strength audio search algorithm. In: Proceedings of the 4th international conference on music information retrieval, pp 7–13
2. Wold E, Blum T, Keislar D, Wheaton J (1996) Content-based classification, search, and retrieval of audio. IEEE Multimedia 3(4):27–36
3. Salehinejad H, Barfett J, Aarabi P, Valaee S, Colak E, Gray B, Dowdell T (2017) A convolutional neural network for search term detection. In: Proceedings of IEEE 28th annual international symposium on personal, indoor, and mobile radio communications (PIMRC), Montreal, QC, pp 1–6
4. Guo G, Li SZ (2003) Content-based audio classification and retrieval by support vector machines. IEEE Trans Neural Netw 14(1):209–215
5. Barrington L, Chan A, Turnbull D, Lanckriet G (2007) Audio information retrieval using semantic similarity. In: Proceedings of IEEE international conference on acoustics, speech and signal processing (ICASSP), vol 2, pp 725–728
6. https://catalog.ldc.upenn.edu/LDC93S1
7. https://catalog.ldc.upenn.edu/LDC2011S02
8. http://www.ldcil.org/resourcesSpeechCorp.aspx
9. http://ice-corpora.net/ice/
10. http://aflat.org/ or http://www.meraka.org.za/lwazi
11. http://catalog.elra.info/index.php?cPath=37
12. https://catalog.ldc.upenn.edu/ldc96s35
13. https://catalog.ldc.upenn.edu/ldc93s6a
14. https://catalog.ldc.upenn.edu/ldc94s13a
15. https://catalog.ldc.upenn.edu/LDC2004S13
16. https://ocw.mit.edu/courses/audio-video-courses/
17. https://www.nist.gov/itl/iad/mig/open-keyword-search-evaluation
18. http://www.multimediaeval.org/about/
19. http://www.multimediaeval.org/mediaeval2011/SWS2011/
20. Metze F, Rajput N, Anguera X, et al (2012) The spoken web search task at Mediaeval 2011. In: IEEE international conference on acoustics, speech and signal processing (ICASSP), pp 5165–5168
21. Metze F, Anguera X, Barnard E, et al (2013) The spoken web search task at Mediaeval 2012. In: ICASSP, pp 8121–8125
22. http://www.multimediaeval.org/mediaeval2014/quesst2014/
23. https://iberspeech2016.inesc-id.pt/index.php/albayzin-evaluation/
24. https://www.iarpa.gov/index.php/research-programs/babel/

25. Alumäe T, Karakos D, Hartmann W, Hsiao R, Zhang L, Nguyen L, Tsakalidis S, Schwartz, R (2017) The 2016 BBN Georgian telephone speech keyword spotting system. In: ICASSP, pp 5755–5759
26. Gupta V, Ajmera J, Kumar A, Verma A, (2011) A language independent approach to audio search. In: Proceedings of INTERSPEECH, pp 1125–01128
27. Manning CD, Raghavan P, Schütze H (2008) Introduction to information retrieval. Cambridge University Press, Cambridge. https://nlp.stanford.edu/IR-book/html/htmledition/evaluation-of-ranked-retrieval-results-1.html
28. Young S, Kershaw D, Odell J, Ollason D, Valtchev V, Woodland P (1999) The HTK book. Entropic Ltd., Cambridge
29. Fiscus JG, Ajot J, Garofolo JS, Doddingtion G (2007) Results of the 2006 spoken term detection evaluation. In: Proceedings of ACM SIGIR workshop on searching spontaneous conversational. Citeseer, pp 51–55
30. Anguera X, Fuentes LR, Buzo A, Metze F, Szoke I, Penagarikano M (2015) QUESST2014: evaluating query-by-example speech search in a zero-resource setting with real-life queries. In: Proceedings of ICASSP, pp 5833–5837
31. Rodriguez-Fuentes LJ, Penagarikano M (2013) MediaEval 2013 spoken web search task: system performance measures, Tech. Rep., Software Technologies Working Group, University of the Basque Country UPV/EHU. http://gtts.ehu.es/gtts/NT/fulltext/rodriguezmediaeval13.pdf
32. O'Shaughnessy D (1987) Speech communication: human and machine. Universities Press, Hyderabad
33. Vasudev D, Gangashetty SV, Anish Babu KK, Riyas KS, (2015) Query-by-example spoken term detection using Bessel features. In: IEEE international conference on signal processing, informatics, communication and energy systems (SPICES'15), pp 1–4. https://doi.org/10.1109/SPICES.2015.7091361
34. Makhoul J (1975) Linear prediction: a tutorial review. Proc IEEE 63(4):561–580
35. Rabiner L, Juang BH, Yegnanarayana B (2009) Fundamentals of speech recognition. Pearson, New Delhi
36. Hermansky H (1990) Perceptual linear predictive (PLP) analysis of speech. J Acoust Soc Am 87(4):1738–1752
37. Thomas S, Ganapathy S, Hermansky H (2008) Recognition of reverberant speech using frequency domain linear prediction. IEEE Signal Process Lett 15:681–684
38. Barnwal S, Sahni K, Singh R, Raj B (2012) Spectrographic seam patterns for discriminative word spotting. In: Proceedings of ICASSP, pp 4725–4728
39. Ezzat T, Poggio T (2008) Discriminative word-spotting using ordered spectro-temporal patch features. In: Proceedings of SAPA workshop, in INTERSPEECH, pp 35–40
40. James DA, Young SJ (1994) A fast lattice-based approach to vocabulary independent word spotting. In: Proceedings of ICASSP, pp 337–380
41. Yu P, Seide F (2004) A hybrid word/phoneme based approach for improved vocabulary-independent search in spontaneous speech. In: INTERSPEECH-2004, pp 293–296
42. Mamou J, Mass Y, Ramabhadran B, Sznajder B (2008) Combination of multiple speech transcription methods for vocabulary independent search. In: Searching spontaneous conversational speech workshop, SIGIR'08, pp 20–27
43. Hori T, Hetherington IL, Hazen TJ, Glass JR (2007) Open-vocabulary spoken utterance retrieval using confusion networks. In: Proceedings of ICASSP, pp 73–76
44. Parada C, Sethy A, Ramabhadran B (2009) Query-by-example spoken term detection for OOV terms. In Proceedings of ASRU, pp 404–409
45. Mangu L, Brill E, Stolcke A (2000) Finding consensus in speech recognition: word error minimization and other applications of confusion networks. In Proceedings of computer speech and language, pp 373–400
46. Muscariello A, Gravier G, Bimbot F, Rennes I, Atlantique B (2009) Audio keyword extraction by unsupervised word discovery. In: INTERSPEECH, pp 2843–2846

47. Chan C, Lee L-S (2010) Unsupervised spoken-term detection with spoken queries using segment-based dynamic time warping. In: INTERSPEECH, 26–30 September 2010, pp 693–696

48. Chan C, Lee L-S (2011) Integrating frame-based and segment-based dynamic time warping for unsupervised spoken term detection with spoken queries. In: Proceedings of ICASSP, pp 5652–5655

49. Anguera X, Ferrarons M (2013) Memory efficient subsequence DTW for query-by-example spoken term detection. In: IEEE international conference on multimedia and expo (ICME), pp 1–6

50. Madhavi MC, Patil HA (2016) Modification in sequential dynamic time warping for fast computation of query-by-example spoken term detection task. In: Proceedings of international conference on signal processing and communications (SPCOM), Bangalore, pp 1–5

51. Bridle JS (1973) An efficient elastic-template method for detecting given words in running speech. In: British acoustical society spring meeting, pp 1–4

52. Higgins AL, Wohlford RE, Bahler LG (1986) Keyword recognition system using template-concatenation model, European Patent Application, Publication No. 0 177 854

53. Rose RC, Paul DB (1990) A hidden Markov model based keyword recognition system. In: Proceedings of ICASSP, vol 1, pp 129–132

54. Weintraub M (1993) Keyword-spotting using SRI's DECIPHER large-vocabulary speech-recognition system. In: Proceedings of ICASSP, vol 2, pp 463–466

55. Weintraub M (1995) LVSCR log-likelihood ratio scoring for keyword spotting. In: Proceedings of ICASSP, vol 1, pp 297–300

56. Iso K, Watanabe T (1990) Speaker-independent word recognition using a neural prediction model. In: Proceedings of ICASSP, pp 441–444

57. Suhardi S, Felbaum K (1997) Wordspotting using a predictive neural model for the telephone speech corpus. In: Proceedings of ICASSP, vol 2, pp 915–918

58. Thambiratnam K, Sridharan S (2005) Dynamic match phone-lattice searches for very fast and accurate unrestricted vocabulary keyword spotting. In: Proceedings of ICASSP, vol 1, pp 465–468

59. Garcia A, Gish H (2006) Keyword spotting of arbitrary words using minimal speech resources. In: Proceedings of ICASSP, vol 1, pp 949–952

60. Audhkhasi K, Verma A (2007) Keyword spotting using modified minimum edit distance measure. In: Proceedings of ICASSP, vol 4, pp 929–932

61. Schwarz P, Matejka P, Burget L, Glembek O (2003) Phoneme recognizer based on long temporal context. Speech Processing Group, Faculty of Information Technology, Brno University of Technology [Online]. Available: http://speech.fit.vutbr.cz/en/software

62. Xin L, Wang BX (2001) Utterance verification for spontaneous Mandarin speech keyword spotting. In: IEEE international conference on Info-tech and Info-net (ICII), vol 3, pp 397–401

63. Ou J, Chen K, Wang X, Li Z (2001) Utterance verification of short keywords using hybrid neural-network/HMM approach. In: IEEE international conference on Info-tech and Info-net (ICII), vol 2, pp 671–676

64. Thambiratnam K, Sridharan S (2003) Isolated word verification using cohort word-level verification. In: EUROSPEECH, pp 905–908

65. Smith G, Murase H, Kashino K (1998) Quick audio retrieval using active search. In: Proceedings of ICASSP, vol 6, pp 3777–3780

66. Wallance R, Vogt R, Sridharan S (2007) A phonetic search approach to the 2006 NIST spoken term detection evaluation. In: INTERSPEECH, pp 2385–2388

67. Thambiratnam K, Sridharan S (2007) Rapid yet accurate speech indexing using dynamic match lattice spotting. IEEE Trans Audio Speech Lang Process 15(1):346–357

68. Pinto J, Szoke I, Prasanna SRM, Hermansky H (2008) Fast automatic spoken term detection from sequence of phonemes. In: Proceedings of 31st annual international ACM SIGIR'08 conference, pp 28–33

69. Wallance R, Vogt R, Sridharan S (2009) Spoken term detection using fast phonetic decoding. In: Proceedings of ICASSP, pp 4881–4884

70. Lin H, Stupakov A, Bilmes J (2008) Spoken keyword spotting via multi-lattice alignment. In: INTERSPEECH, pp 2191–2194
71. Vergyri D, Shafran I, Stolcke A, Gadde RR, Akbacak M, Roark B, Wang W (2007) The SRI/OGI 2006 spoken term detection system. In: Proceedings of INTERSPEECH, pp 2393–2396
72. Chen G, Parada C, Heigold G (2014) Small-footprint keyword spotting using deep neural networks. In: Proceedings of ICASSP, pp 4087–4091
73. Sainath TN, Parada C (2015) Convolutional neural networks for small-footprint keyword spotting. In: INTERSPEECH, pp 1478–1482
74. Lim H, Kim Y, Kim Y, Kim H (2017) CNN-based bottleneck feature for noise robust query-by-example spoken term detection. In: Proceedings of Asia-Pacific signal and information processing association annual summit and conference (APSIPA ASC), Kuala Lumpur, pp 1278–1281
75. Zhang Y (2013) Unsupervised speech processing with applications to query-by-example spoken term detection, Ph.D. Thesis, Department of Electrical Engineering and Computer Science, Massachusetts Institute of Technology
76. Mantena G, Achanta S, Prahallad K (2014) Query-by-example spoken term detection using frequency domain linear prediction and non-segmental dynamic time warping. IEEE/ACM Trans Audio Speech Lang Process 22(5):946–955
77. Hazen TJ, Shen W, White C (2009) Query-by-example spoken term detection using phonetic posteriorgram templates. In: Proceedings of IEEE workshop on automatic speech recognition & understanding (ASRU), pp 421–426
78. George B, Yegnanarayana B (2014) Unsupervised query-by-example spoken term detection using segment-based bag of acoustic words. In: Proceedings of ICASSP, pp 7183–7187
79. Zhang Y, Glass JR (2009) Unsupervised spoken keyword spotting via segmental DTW on Gaussian posteriorgrams. In: Proceedings of ASRU, pp 398–403
80. Obara M, Moriya M, Konno R, Kojima K, Tanaka K, Lee S, Itoh Y (2017) Acceleration for query-by-example using posteriorgram of deep neural network. In: Proceedings of APSIPA ASC, Kuala Lumpur, pp 1565–1569
81. Ram D, Asaei A, Bourlard H (2018) Sparse subspace modeling for query by example spoken term detection. In: IEEE/ACM transactions on audio, speech, and language processing, vol 26, no 6, pp 1130–1143
82. Wang H, Lee T, Leung C (2011) Unsupervised spoken term detection with acoustic segment model. In: International conference on speech database and assessments (Oriental COCOSDA), pp 106–111
83. Dumpala SH, Raju Alluri KNRK, Suryakanth VG, Uppala AKV (2015) Analysis of constraints on segmental DTW for the task of query-by-example spoken term detection. In: Annual IEEE India conference (INDICON), New Delhi, pp 1–6
84. Nagarajan T, Murthy HA, Hegde RM (2003) Segmentation of speech into syllable-like units. In: EUROSPEECH, pp 2893–2896
85. Madikeri SR, Murthy HA (2012) Acoustic segmentation using group delay functions and its relevance to spoken keyword spotting. In: Text speech and dialogue. Springer, Heidelberg, pp 496–504
86. Karthik Pandia DS, Saranya MS, Murthy HA (2016) A fast query-by-example spoken term detection for zero resource languages. In: IEEE SPCOM'16, pp 1–5
87. Sankar R, Jain A, Deepak KT, Vikram CM, Prasanna SRM (2016) Spoken term detection from continuous speech using ANN posteriors and image processing techniques. In: IEEE 22nd national conference on communication (NCC), pp 1–6
88. Madhavi MC, Patil H (2017) Partial matching and search space reduction for QbE-STD. Comput Speech Lang 45:58–82

Index